SODIUM CHANNELS AND NEURONAL HYPEREXCITABILITY

The Novartis Foundation is an international scientific and educational charity (UK Registered Charity No. 313574). Known until September 1997 as the Ciba Foundation, it was established in 1947 by the CIBA company of Basle, which merged with Sandoz in 1996, to form Novartis. The Foundation operates independently in London under English trust law. It was formally opened on 22 June 1949.

The Foundation promotes the study and general knowledge of science and in particular encourages international co-operation in scientific research. To this end, it organizes internationally acclaimed meetings (typically eight symposia and allied open meetings and 15–20 discussion meetings each year) and publishes eight books per year featuring the presented papers and discussions from the symposia. Although primarily an operational rather than a grant-making foundation, it awards bursaries to young scientists to attend the symposia and afterwards work with one of the other participants.

The Foundation's headquarters at 41 Portland Place, London W1N 4BN, provide library facilities, open to graduates in science and allied disciplines. Media relations are fostered by regular press conferences and by articles prepared by the Foundation's Science Writer in Residence. The Foundation offers accommodation and meeting facilities to visiting scientists and their societies.

Information on all Foundation activities can be found at http://www.novartisfound.org.uk

Novartis Foundation Symposium 241

SODIUM CHANNELS AND NEURONAL HYPEREXCITABILITY

2002

JOHN WILEY & SONS, LTD

Published in 2002 by John Wiley & Sons Ltd,
Baffins Lane, Chichester,
West Sussex PO19 1UD, England

National 01243 779777
International (+44) 1243 779777
e-mail (for orders and customer service enquiries): cs-books@wiley.co.uk
Visit our Home Page on http://www.wiley.co.uk
or http://www.wiley.com

Other Wiley Editorial Offices

John Wiley & Sons, Inc., 605 Third Avenue,
New York, NY 10158-0012, USA

WILEY-VCH Verlag GmbH, Pappelallee 3,
D-69469 Weinheim, Germany

Jacaranda Wiley Ltd, 33 Park Road, Milton,
Queensland 4064, Australia

John Wiley & Sons (Asia) Pte Ltd, 2 Clementi Loop #02-01,
Jin Xing Distripark, Singapore 129809

John Wiley & Sons (Canada) Ltd, 22 Worcester Road,
Rexdale, Ontario M9W 1L1, Canada

Novartis Foundation Symposium 241
viii+244 pages, 42 figures, 5 tables

Library of Congress Cataloging-in-Publication Data

Sodium channels and neuronal hyperexcitability / [editors], Gregory Bock, Jamie A. Goode.
p. cm. – (Novartis Foundation symposium ; 241) (Ciba Foundation symposium)
Includes bibliographical references and index.
ISBN 0-471-48530-6 (alk. paper)
1. Sodium channels–Congresses. 2. Molecular neurobiology–Congresses. 3. Nervous system–Diseases–Molecular aspects–Congresses. 4. Neurons–Congresses. I. Bock, Gregory. II. Goode, Jamie. III. Series. IV. Series: Ciba Foundation symposium
QP356.2.S65 2001
612.8′042–dc21 2001046625

British Library Cataloguing in Publication Data

A catalogue record for this book is available from the British Library

ISBN 0 471 48530 6

Typeset in $10\frac{1}{2}$ on $12\frac{1}{2}$ pt Garamond by Dobbie Typesetting Limited, Tavistock, Devon.
Printed and bound in Great Britain by Biddles Ltd, Guildford and King's Lynn.
This book is printed on acid-free paper responsibly manufactured from sustainable forestry, in which at least two trees are planted for each one used for paper production.

Contents

Symposium on Sodium channels and neuronal hyperexcitability, held at the Novartis Foundation, London, 14–16 November 2000

Editors: Gregory Bock (Organizer) and Jamie A. Goode

This symposium is based on a proposal made by John Wood and Stephen Waxman in conjunction with the Yale–University College London Collaboration

Participants

Mark D. Baker Department of Biology, University College London, Medawar Building, Gower Street, London WC1E 6BT, UK

Bruce Bean Department of Neurobiology, Harvard Medical School, 220 Longwood Avenue, Boston, MA 02115, USA

Stuart Bevan Novartis Institute for Medical Science, University College London, Gower Street, London WC1E 6BT, UK

Hugh Bostock Sobell Department of Neurophysiology, Institute of Neurology, Queen Square, London WC1N 3BG, UK

William A. Catterall Department of Pharmacology, F-427 Health Sciences Center, University of Washington, Seattle, WA 98195-7280, USA

Wayne Crill Department of Physiology and Biophysics, University of Washington, 1959 NE Pacific Street, HSB Room G424, Box 357290, Seattle, WA 98195-7290, USA

Ted R. Cummins Department of Neurology, Yale University, New Haven, CT 06520, USA

Michael S. Gold Oral and Craniofacial Biological Sciences, University of Maryland Dental School, 666 W. Baltimore Street, Baltimore, MD 21201, USA

Alan L. Goldin Department of Microbiology & Molecular Genetics, University of California at Irvine, Irvine, CA 92697-4025, USA

Richard Horn Department of Physiology, Institute of Hyperexcitability, Jefferson Medical College, 1020 Locust Street, Philadelphia, PA 19107, USA

Lori L. Isom Department of Pharmacology, The University of Michigan Medical School, 1301 MSRB III, 1150 W. Medical Center Drive, Ann Arbor, MI 48109-0632, USA

Richard Keynes Department of Physiology, University of Cambridge, Downing Street, Cambridge CB2 3EG, UK

Miriam Meisler Department of Human Genetics, University of Michigan Medical School, Ann Arbor, MI 48109-0618, USA

Jeffrey L. Noebels Department of Neurology, Baylor College of Medicine, One Baylor Plaza, Houston, TX 77030, USA

Louis Ptacek Howard Hughes Medical Institute, University of Utah School of Medicine, Room 4425, Salt Lake City, UT 84112-5331, USA

Indira M. Raman Department of Neurobiology and Physiology, Northwestern University, 2153 North Campus Drive, Evanston, Illinois 60208-3520, USA

Michael Segal Harvard Medical School and Brigham & Women's Hospital, Longwood Medical Research Center, 221 Longwood Ave, Boston MA 02115, USA

Nelson Spruston Department of Neurobiology and Physiology, Northwestern University, 2153 N. Campus Drive, Evanston, IL 60208-3520, USA

Gary R. Strichartz Pain Research Center, Harvard Medical School, Anaesthesia Research Laboratories, Medical Research Building 611, Brigham and Women's Hospital, 75 Francis Street, Boston, MA 02115, USA

Stephen Waxman (*Chair*) Department of Neurology, Yale Medical School, 333 Cedar Street, LCI-707, New Haven, CT 06510, USA

John Wood Department of Biology, University College London, Medawar Building, Gower Street, London WC1E 6BT, UK

Chair's introduction: sodium channels and neuronal dysfunction — emerging concepts, converging themes

Stephen G. Waxman

Department of Neurology, Yale Medical School, 333 Cedar Street, LCI-707, New Haven, CT-06510, USA

It is nearly 50 years since Hodgkin and Huxley published their pioneering work on Na^+ channels. And it has been an incredibly productive 50 years. We can no longer talk about *the* Na^+ channel, since we now appreciate the existence of a family of channels, each member apparently designed to function in a different and special way. Although it is almost axiomatic that Na^+ channels endow the nervous system with electrogenicity, we are learning more and more, by the month, about the role of Na^+ channels in electrogenesis, and about the many ways in which the various channels contribute to it. We also now know that Na^+ channels are key players in some neurological diseases, including disorders that reflect neuronal hyperexcitability.

This meeting will focus on an emerging concept, that a multiplicity of molecularly and physiologically distinct Na^+ channels contribute to neuronal hyperexcitability that can produce clinically significant signs and symptoms. This concept, and our meeting which will explore it, reflects the convergence of a number of important themes:

The first theme is channel diversity. We are recognizing an increasing degree of diversity among Na^+ channels. We now understand that at least 10 different genes encode molecularly distinct Na^+ channels. Interestingly, several cell-specific Na^+ channels have been identified, including Na^+ channels that are preferentially expressed within dorsal root ganglion neurons and trigeminal neurons.

The molecular diversity of Na^+ channels appears to be paralleled by diversity in terms of physiological and pharmacological properties, and in terms of functional roles. We are learning that Na^+ channel diversity contributes to the tuning of neurons, so that different types of neurons can sing different songs, or at least produce music in different keys. In some cases the songs of neurons can be discordant, or even shrill. This can contribute to neuropathic pain, which we will discuss during this meeting.

Theme one—channel diversity—has also, almost predictably, created a problem with nomenclature. As the number of Na^+ channels that have been identified has expanded, there have been numerous attempts to name these various channels and the situation has become confusing. To address this issue, Al Goldin has spearheaded an effort to establish a uniform nomenclature of Na^+ channels, a summary of which can be found in Table 1. The new nomenclature will undoubtedly help us as we discuss the multiplicity of sodium channels that contribute to neuronal function and dysfunction.

Theme number two is a very rich biology. Different channels may subserve different roles in different cell types, and possibly under different conditions. We are beginning to understand the complex network of factors and molecules that control or modulate channel expression and function. We are also beginning to understand dynamic aspects of channel expression. Recent work suggests that plasticity of Na^+ channel expression can play an adaptive role in the normal nervous system, but there are also dynamic aspects of channel expression in diseased neurons, and in some cases changes are maladaptive.

The third theme concerns molecular and submolecular analysis. We are learning more and more about structure–function relations at a molecular level; for example, we are beginning to understand the molecular substrates for gating, voltage dependence and channel kinetics. Within α subunits, this includes understanding the importance of single amino acids or small subdomains in conferring specific physiological or pharmacological properties on some types of channels. β subunits are multiple and diverse, with an ensemble of functions and behaviours so rich that one investigator in this room has suggested that α subunits may be accessories to β subunits!

The fourth theme is pathophysiology. At this meeting we will be talking about the potential contributions of Na^+ channels to neuronal hyperexcitability. We will see that anomolously expressed Na^+ channels can contribute to altered excitability by their function as inward current generators at a given time. We will also see that sodium channels can contribute to pathophysiology when they are expressed in acquired channelopathies in abnormal ways. Abnormal Na^+ channel expression has clearly been demonstrated to contribute to neuropathic pain, and it may contribute to symptom production in multiple sclerosis. We will of course be hearing a lot about genetics: new and in many cases instructive mutations of Na^+ channels have been identified. Some of these mutations cause specific, well-defined syndromes, which can teach us important things about channels and their functions. We are only just beginning to appreciate the extent of the genetic channelopathies; an increasing spectrum of neurological disorders that involve mutated channels is being identified, one of which is epilepsy.

The fifth and final theme for this meeting concerns the pharmacology of Na^+ channels. There is a molecular pharmacology of Na^+ channels on the horizon.

TABLE 1 Mammalian sodium channel α subunits

Type	*Alias*	*Genbank number*[a]	*Gene symbol*	*Chromosomal location*	*Splice variants*	*Alias*
$Na_v1.1$	rat I HBSCI GPBI SCN1A	X03638 (R) X65362 (H) AF003372 (GP)	SCN1A	Mouse 2 (36) Human 2q24	$Na_v1.1a$	rat Ia
$Na_v1.2$	rat II HBSCII HBA	X03639 (R) X65361 (H) M94055 (H)	SCN2A	Mouse 2 (36) Human 2q23-24	$Na_v1.2a$	rat IIA
$Na_v1.3$	rat III	Y00766 (R)	SCN3A	Mouse 2 (36) Human 2q24	$Na_v1.3a$ $Na_v1.3b$	rat IIIa rat IIIb
$Na_v1.4$	SkM1, μ1	M26643 (R) M81758 (H)	SCN4A	Mouse 11 (64) Human 17q23-25		
$Na_v1.5$	SkM2 H1	M27902 (R) M77235 (H)	SCN5A	Mouse 9 (70) Human 3p21		
$Na_v1.6$	NaCh6 PN4 Scn8a CerIII	L39018 (R) AF049239 (R) AF049240 (R) U26707 (M) AF049617 (M) AF050736 (H) AF003373 (GP)	SCN8A	Mouse 15 (64) Human 12q13	$Na_v1.6a$	PN4a
$Na_v1.7$	NaS hNE-Na PN1	U35238 (R) X82835 (H) AF000368 (R) U79568 (R)	SCN9A	Mouse 2 (36) 2 Human 2q24		
$Na_v1.8$	SNS PN3 NaNG	X92184 (R) U53833 (R) Y09108 (M) U60590 (D)	SCN10A	Mouse 9 (67) [26] Human 3p22-24		
$Na_v1.9$	NaN SNS2 PN5 NaT SCN12A	AF059030 (R) AJ237852 (R) AF118044 (M) AB031389 (M) AF126739 (H) AF109737 (H)	SCN11A	Mouse 9 [11,33] Human 3p21-24	$Na_v1.9a$	SCN12
Na_x	$Na_v2.1$ Na-G SCL11 $Na_v2.3$	M91556 (H) M96578 (R) Y09164 (R) L36179 (M)	SCN7A (SCN6A)[b]	Mouse 2 (41) Human 2q21-23		

[a]The letter in parentheses after each accession number indicates the species of origin for the sequence, as follows: H, human; R, rat; M, mouse; GP, guinea pig; D, dog.
[b]This gene was originally assigned symbols SCN6A and SCN7A, which were mapped in human and mouse, respectively. The two most likely represent the same gene, and the SCN6A symbol will probably be deleted.
Reproduced with permission from Goldin et al (2000).

The past decade has given us new drugs and toxins that modulate channels, and we are gaining an increased understanding of their mechanism of action. We are beginning to define the substrates for the actions of channel modulating drugs and toxins at an atomic level.

I hope that during this meeting we can pursue several overarching questions. One is, how do different subtypes of Na^+ channels contribute to neuronal hyperexcitability? Second, is it possible to target the different Na^+ channel subtypes, in ways that might control or modulate neuronal hyperexcitability? Third, can we exploit, for therapeutic advances, the selective expression patterns of some types of channels? Obviously, we are not going to solve these questions here, but I hope that we can at least think about them and possibly define some strategies or paradigms.

This meeting is in many ways an outgrowth of the Yale–London Collaboration, and I am indebted both to Yale and to University College London for encouragement and support. It is also a pleasure to acknowledge John Wood, my friend and co-organizer, who played a major role in organizing this meeting.

Again, I'd like to welcome everyone and thank you for coming. I know we will have an interesting week.

Reference

Goldin AL, Barchi RL, Caldwell JH et al 2000 Nomenclature of voltage-gated sodium channels. Neuron 28:365–368

Studies of multimodal gating of the sodium channel

Richard D. Keynes

Physiological Laboratory, University of Cambridge, Downing Street, Cambridge CB2 3EG, UK

Abstract. Chandler and Meves found that in squid axons perfused with NaF a small flow of Na^+ ions persisted in the inactivated state, and that the Na^+ channel therefore has more than one open state. Studies by Correa and Bezanilla on single patches in squid axons showed that such steady currents arose from reopening of the channel at a relatively low frequency. Currents with comparable properties are generated in mammalian brain cells and elsewhere. The existence of a third mode of gating was established by Patlak and Ortiz when they showed that in frog muscle fibres there were occasionally quite large bursts of late openings. Again, similar behaviour has been observed in other types of muscle and in brain cells. It is suggested that the voltage gating of all ionic channels involves a screw-helical mechanism, operating in steps each transferring unit charge. For segment S4 in domain IV of Na^+ channels, three charges have to be transferred to reach the initial open state, and a fourth for fast inactivation to take place. The single late openings in the inactivated steady state may be explained by the transfer of a fifth charge in IVS4, while the larger bursts of reopening involve a modulation of the mechanism of fast inactivation.

2002 Sodium channels and neuronal hyperexcitability. Wiley, Chichester (Novartis Foundation Symposium 241) p 5–20

As was recalled by Keynes (1994a), it was first observed by Chandler & Meves (1970) that in squid axons perfused with NaF a small flow of Na^+ ions persisted in the inactivated steady state. Their tentative conclusion based on the Hodgkin–Huxley model was to suppose that the inactivation parameter h was the sum of two components h_1 and h_2, where h_1 predominated at negative potentials, and h_2 predominated at positive ones. This proposition predicted the existence of one type of Na^+ conductance that increased transiently with depolarization as in the Hodgkin–Huxley system, and a second type that persisted with depolarization to give a steady low level of conductance. Combined with m^3 kinetics the idea fitted well with the experimental data.

It was subsequently suggested that the phenomenon might alternatively be explained by the existence in parallel of two types of channel with different characteristics. But studies by Correa & Bezanilla (1994) in patch-clamped squid

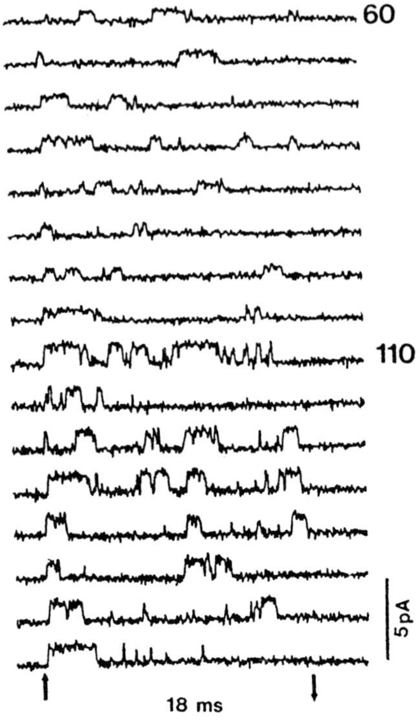

FIG. 1. Reopening of a single channel from a squid axon at potentials of +60 mV (upper block) and +110 mV (lower block), recorded at 5 °C in 270 Na/535 Na. Onset and release of the 18 ms depolarizations are indicated by the arrows at the bottom. Holding potential −100 mV. From Correa & Bezanilla (1994).

axons showed elegantly and conclusively, as may be seen in Fig. 1, that in a patch containing a single channel the steady current arose from the arrival of the channel a short while after its initial opening, followed by its fast inactivation, at a state in which it continued to reopen at a relatively low frequency. It was thus clear that each channel had two different routes to an open state.

It was suggested mistakenly by Hille (1992) that incomplete inactivation of this type 'may be unique to squid axons', but Alzheimer et al (1993a,b) demonstrated the development of a persistent Na^+ current in pyramidal neurons from rat and cat sensorimotor cortex, and suggested that a switching between channels exhibiting different degrees of modal gating might be of importance in early development. Other examples of this type of bimodal gating have since come to light.

Measurements have been reported by Keynes & Meves (1993) of the probability functions both for the initial opening of the Na^+ channels in squid axons, and for the reopenings that generate the late current in the inactivated steady state. Plots

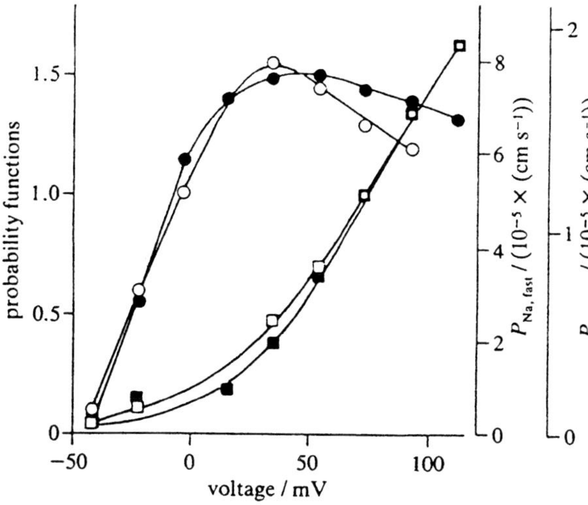

FIG. 2. The probability functions PF_{peak} for initial opening (open circles) and PF_{ss} for opening in the inactivated steady state (open squares), plotted against test potential, for a squid axon bathed in 514 mM Na+16 nM TTX and dialysed with 350 mM NaF. Temperature 5 °C. Holding potential −80 mV. The solid symbols show corresponding values of the macroscopic permeability coefficients. From Keynes & Meves (1993).

against potential of the probability functions PF_{peak} for initial opening, and PF_{ss} for reopening in the steady state, are shown in Fig. 2. It will be seen that whereas PF_{peak} rises to an early peak as expected, and then declines slightly, PF_{ss} rises with a delay and above 50 mV continues upwards on a straight line. Analysis of the data showed that just as arrival at the initial open state of the channel has been shown by Keynes & Elinder (1999) to involve a series of steps each carrying close to one electronic charge. The late reopenings also require the transfer of slightly less than one charge, though with an equilibrium potential shifted positively by about 95 mV as compared with that for PF_{peak}.

The next step in my argument must be to consider the nature of the mechanism responsible for these two different types of gating. There is now general agreement that the voltage sensors of the system are the positive charges carried by the S4 transmembrane segments in each of the four domains of every channel. The total transfer of charge needed to bring about the normal opening of both K^+ and Na^+ channels is close to 12 electronic units, and thus three in each S4 segment. A vitally important characteristic of the S4 segments is that the great majority of the positively charged lysine and arginine residues exhibit a one in three spacing with two uncharged residues situated between each of them. This spacing is perfectly conserved across the entire animal kingdom in all voltage-gated ion channels, whether they are selective for Na^+, K^+ or Ca^{2+}. But the total number of positive charges carried by a single segment varies from four to eight between the several domains. In Na^+ channels there are almost always four in domain I, five in domains II and III, and eight in domain IV.

Another less well recognized structural feature of all voltage-gated ion channels, to which attention was recently drawn by Keynes & Elinder (1999), is the near perfect conservation of the location of negatively charged glutamate and aspartate residues in segments S2 and S3. Thus in S3 there is an aspartate residue located six places from the inner end of the segment, and in S2 invariably a glutamate three places nearer the centre of the segment. The position in every domain of an outermost negative charge, or sometimes of a pair of charges, is conserved with less precision, and they may either be glutamate or aspartate, located either on S2 or S3. It would seem that the prime function of the triplet of fixed negative charges is to pair up with three of the mobile positive charges and so stabilize their movements in discrete steps across the membrane. Three-dimensional models have shown that if S4 is appropriately tilted relative to S2 and S3 lying parallel to one another, ion pairs can readily be formed. The precision of their location in every voltage gated ion channel in the world, since sodium channels first evolved around 550 million years ago, confirms that their role is of central importance.

The explanation originally put forward by Catterall (1986) and by Guy & Seetharamulu (1986) for the one in three spacing of the positive charges was that their movements take place in a screw-helical fashion, which meant that a 60° twist of the S4 α-helix from a position in which its positive charges were paired up with fixed negative ones would move it out 0.45 nm, bringing each positive charge into the position previously occupied by its neighbour, where it would automatically find a negative charge with which to pair. However, an obvious objection to the screw-helical theory was that the total number of positive charges carried by the four S4 segments was appreciably greater than the number of negative charges carried by the other five segments. It was also hard to believe that in the open state the outermost arginines of S4 would project as far as 1.45 nm into the external aqueous phase. So for some time the theory lost its initial popularity.

Taking advantage of the naturally occurring mutation R1448C of segment IVS4 in human skeletal muscle Na^+ channels in which cysteine replaces the outermost arginine, Yang & Horn (1995) introduced a technique for testing the accessibility of the cysteine residue to hydrophilic methanethiosulfonate (MTS) and methanethiosulfonate-ethyltrimethylammonium (MTSET) reagents, which demonstrated clearly that it was indeed moved into the external aqueous phase by depolarization. This technique was subsequently extended by Yang et al (1997) to carry out more extensive observations on the accessibility of several of the positive charges to either the aqueous external or internal environments when they were mutated to cysteine. Similar observations were made on Shaker K^+ channels by Larsson et al (1996). An important conclusion from this work was that the effective width of the low dielectric constant portion of the membrane across which the electric field acts to pull the positive charges outwards or inwards must be substantially less than the 3 nm that corresponds to a membrane capacity of 1 μF.cm^{-2}. It is nevertheless wide enough to house the three well conserved negative charges now shown to be located on S2 and S3 in every domain.

Fig. 3 shows a strictly diagrammatic representation of the screw-helical mechanism put forward by Keynes & Elinder (1999) to accommodate the MTSET accessibility data of Larsson et al (1996) and of Yang et al (1997), as applied to segment IVS4 of a human skeletal muscle Na^+ channel. In the strongly hyperpolarized closed state, five of the positive charges project inwards into the aqueous phase, while in the open state depolarized to around 0 mV three of them are exposed to the external aqueous phase, and only two still project inwards. Another positive charge is transferred outwards in order to inactivate the channel, and the last and fifth does so for the final reopening in the inactivated steady state. It should be noted that the data of Keynes (1994a) indicate that hyperpolarization to −180 mV is needed to complete the inward transfer of the gating charge, while depolarization to +100 mV is needed to reopen an appreciable number of channels. The more restricted accessibility of the channels reported by Yang et al (1997) and Horn (2000) refers to a somewhat narrower voltage range.

I should make clear at this point that the relatively straightforward picture of the screw-helical theory shown in Fig. 3, in which α-helix S4 both rotates and moves outwards, conflicts seriously with the views of Bezanilla et al (2001). Bezanilla currently maintains that the four outermost positive charges reside in long crevices with ends too narrow to fit MTS reagents, but large enough to allow protons to have access to them. On depolarization, the S4 segment is claimed to rotate with little or no outward translation so that the charges which are initially in an internally exposed crevice move to an externally connected one. It has, however, been pointed out by Horn (2000) that the attachment of large fluorophores to measure such movements of subunits within proteins yields results that are hard to interpret, and there is not yet agreement as to exactly what distances are moved. Most recently, Gandhi et al (2000) have concluded from fluorescence scans of a voltage-gated K^+ channel that the rotation of the S4 segment is accompanied by an appreciable outward movement. I have to confess that I am unable to understand where precisely Bezanilla's crevices are in fact located, the absence of any hard anatomical evidence in support of their existence being for me the major weakness in his arguments. It is also unclear to me how this mechanism is supposed to transfer unit charges in discrete steps, for which there is

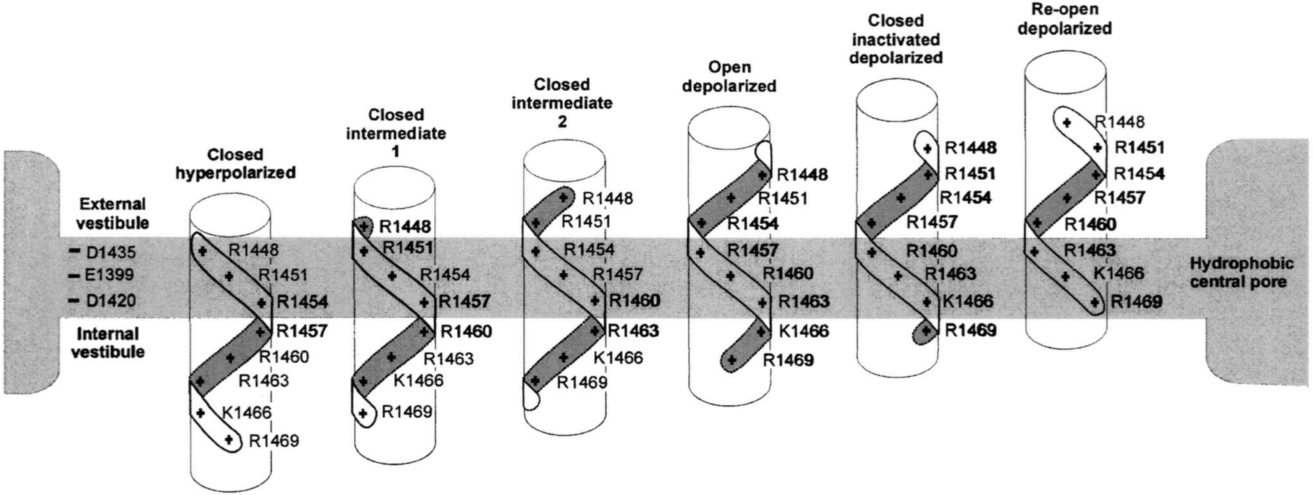

FIG. 3. A strictly diagrammatic representation of the screw-helical outward movement of the positive charges carried by segment IVS4 in a human muscle Na^+ channel. Each outward step transfers one electronic charge from the interior of the cell to the external solution. The three negative charges shown on the left occupy fixed positions on IVS2 and IVS3, and form salt bridges with three of the mobile positive charges. After the version for a Shaker K^+ channel of Keynes & Elinder (1999).

incontrovertible evidence in the operation of Na^+ channels; nor does it appear to take account of the implications of the conspicuously good conservation of the location of the negative charges on S2 and S3 that I have described.

However, I must also admit that my modified version of the screw-helical hypothesis is open to serious objection in that the outward movements of more than 2 nm in the IVS4 segment of a Na^+ channel shown in Fig. 3 may be hard to reconcile with the shortness of the external links, some of which contain only four residues between the outer ends of segments S3 and S4. I can only suggest as a possible solution to this difficulty that the emergence of the ends of the S4 segments into the external aqueous phase may be accompanied by an instantaneous collapse of their α-helical structure.

Although the current generated by reopening of channels in the inactivated steady state may well serve for a fine tuning of the excitability of neurons in the course of development, it could not readily be responsible for the explosive increases in excitability which I take to be the main topic of interest for this meeting. It is therefore necessary to look for a third mode of gating of the Na^+ channel. From the extensive literature on the involvement of voltage-gated ion channels in hereditary disease recently summarized by Lehmann-Horn & Jurkat-Rott (1999), the most relevant part of the system to examine would evidently be the mechanism of fast inactivation, for whereas there seem to be few if any hereditary mutations that directly increase excitability, there are many that have been shown to operate indirectly by drastically slowing down fast inactivation. Thus in Fig. 4 you will note that the majority of the symbols are clustered around segment IVS4 itself, and the internal link between domains III and IV where the so-called inactivation particle is located.

A relevant phenomenon that has never been observed in a squid axon, but was first reported for patches from frog muscle fibres by Patlak & Ortiz (1986), and is illustrated in Fig. 5, is the appearance in addition to the single late openings seen in Fig. 1 of occasional bursts of 10 s to 100 s of openings.

This occurrence of multiple gating modes in a single population of channels has been reported in other types of muscle and in brain cells, and is characterized by an alternation between periods of inactivation at two different rates, one 10 times slower than the other. It has been suggested that the two modes represent two conformations of the α subunit, one of which can be stabilized either by hyperpolarization or, as has been shown by Ma et al (1997), by modulation of the β_1-units by G proteins. A third mode of gating of this kind therefore seems to be a good candidate for explaining large increases in excitability.

Nevertheless, while I welcome the recent though belated admission of the truth of the proposition for which I have long fought with very few allies (Keynes 1994b), that fast inactivation of the Na^+ channel is indeed a voltage-dependent process, it has to be recognized that a full understanding of the detailed

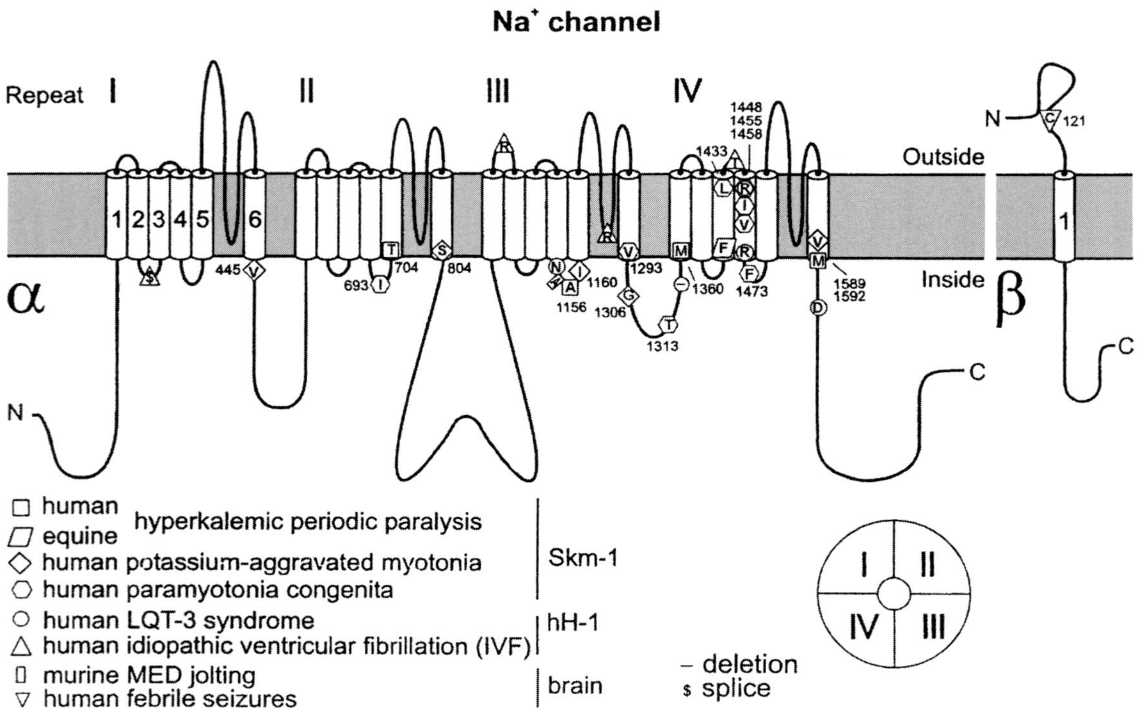

FIG. 4. The transmembrane topology of some mammalian muscle Na^+ channels showing the points where single site mutations interfere with fast inactivation and give rise to the hereditary diseases listed. From Lehmann-Horn & Jurkat-Rott (1999).

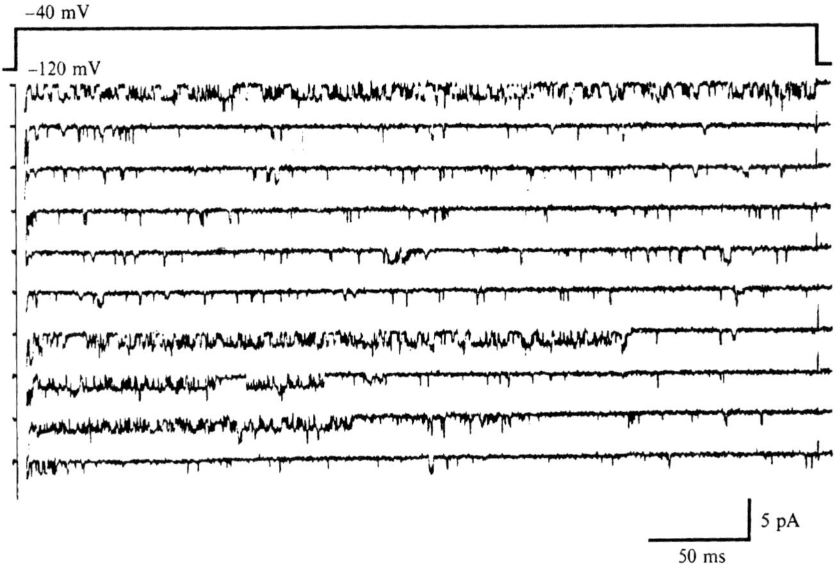

FIG. 5. Late Na^+ currents during 10 sequential pulses to −40 mV applied to a cell-attached patch from a frog muscle fibre. The peaks of initial current reaching −45 pA at the start of each trace have been blanked out. Holding potential −120 mV. Temperature 10 °C. From Patlak & Ortiz (1986).

mechanism of coupling of inactivation of the Na^+ channel is at present no nearer to our grasp than it is for its activation and opening.

References

Alzheimer C, Schwindt PC, Crill WE 1993a Modal gating of Na^+ channels as a mechanism of persistent Na^+ current in pyramidal neurons from rat and cat sensorimotor cortex. J Neurosci 13:660–673

Alzheimer C, Schwindt PC, Crill WE 1993b Postnatal development of a persistent Na^+ current in pyramidal neurons from rat sensorimotor cortex. J Neurophysiol 69:290–292

Bezanilla F, Starace D, Cha A, Selvin PR 2001 Structural underpinnings of voltage sensing in potassium channels. J Gen Physiol, in press (Soc Gen Physiol Symp 54, Massachusetts, 2000)

Catterall WA 1986 Voltage-dependent gating of sodium channels: correlating structure and function. Trends Neurosci 9:7–10

Chandler WK, Meves H 1970 Evidence for two types of sodium conductance in axons perfused with sodium fluoride solution. J Physiol 211:653–678

Correa AM, Bezanilla F 1994 Gating of the squid sodium channel at positive potentials: II. Single channels reveal two open states. Biophys J 66:1864–1878

Gandhi CS, Loots E, Isacoff EY 2000 Reconstructing voltage sensor–pore interaction from a fluorescence scan of a voltage-gated K^+ channel. Neuron 27:585–595

Guy HR, Seetharamulu P 1986 Molecular model of the action potential sodium channel. Proc Natl Acad Sci USA 83:508–512

Hille B 1992 Ionic channels of excitable membranes, 2nd edn. Sinauer Associates, Sunderland, MA, USA
Horn R 2000 A new twist in the saga of charge movement in voltage-dependent ion channels. Neuron 25:511–514
Keynes RD 1994a Bimodal gating of the Na^+ channel. Trends Neurosci 17:58–61
Keynes RD 1994b The kinetics of voltage-gated ion channels. Q Rev Biophys 27:339–434
Keynes RD, Meves H 1993 Properties of the voltage sensor for the opening and closing of the sodium channels in the squid giant axon. Proc R Soc Lond B Biol Sci 253:61–68
Keynes RD, Elinder F 1999 The screw-helical voltage gating of ion channels. Proc R Soc Lond B Biol Sci 266:843–852
Larsson HP, Baker OS, Dhillon DS, Isacoff EY 1996 Transmembrane movement of the Shaker K^+ channel S4. Neuron 16:387–397
Lehmann-Horn F, Jurkat-Rott K 1999 Voltage-gated ion channels and hereditary disease. Physiol Rev 79:1317–1372
Ma JY, Catterall WA, Scheuer T 1997 Persistent sodium currents through brain sodium channels induced by G protein $\beta\gamma$ subunits. Neuron 19:443–452
Patlak JB, Ortiz M 1986 Two modes of gating during late Na^+ channel currents in frog sartorius muscle. J Gen Physiol 87:305–326
Yang N, Horn R 1995 Evidence for voltage-dependent S4 movement in sodium channels. Neuron 15:213–218
Yang N, George AL Jr, Horn R 1997 Probing the outer vestibule of a sodium channel voltage sensor. Biophys J 73:2260–2268

DISCUSSION

Horn: How confident are you that you can calculate time constants from slopes of your gating current? Are you sure that you are looking at single transitions in a kinetic scheme that probably has in the order of at least 60 kinetic states?

Keynes: I can be confident that I am looking at a mixture of them, but I quite agree that they are entering at different stages. If you look at Figs 1 and 5 of Keynes & Elinder (1998a) in which I showed the kinetics of the gating current as a whole, it turns out that when I am measuring those relaxing phases, both the initial events are just about over: we are therefore looking mainly at the relaxation of the third and final component that opens the channel.

Horn: One pleasing aspect was that the lines were straight over quite a long range.

Keynes: They are impressively straight.

Horn: Over the years I have had many discussions with Clay Armstrong, and he believed that there were only a small number of charges that underlie the gating, compared to some of the initial estimates from the K^+ channels. For example, from his measurements of the gating in K^+ and Na^+ channels in squid axon, he thought that there were perhaps six charges underlying the gating, whereas some of the recent data from the Shaker mouse K^+ and Na^+ channels suggest 12–13 charges underlying the whole gating. Eventually he conceded that there had to be many more charges involved, but the difficulty is finding out this sort of information from a macroscopic measurement. One can often underestimate the number of charges, because you are looking at what you can see by fitting exponentials, which is often very much less than what is there. I wouldn't go so far as saying that it is an epiphenomenon, but it is a generalization of the underlying events. I don't think it is possible to pull out all of the exponential components that are actually there from a gating current measurement, or from an ionic current measurement. This is why I am concerned about how confident one can be that there are only single charges carried by those transitions that you see.

Keynes: If there were some that were carrying 2.3 charges, that ought to make the slopes somewhat bigger. In all the plots that I showed there was no trace of a bigger slope. The rising phase is not fitted by a straight line, but it still corresponds to no more than unit charges.

Horn: What about the possibility that the 2.3 occurs in the transition from the rising part of your curve to the falling part.

Keynes: As you may have seen in the plots, I quite often gave up fitting them, because the ratio of the large to the small meant that it was not possible to do an accurate fit. I don't think we have room to explain all of the charges. As you say, one of the facts that everyone agrees on is that it is about 12 charges altogether that are involved, but to get as many as this you have to push the membrane right back to −180 mV for the extreme hyperpolarization, and we also go up to +50 mV or more. It seems to me that if there really were any single movements carrying 2.3, they would show up. The only place they could show up would be at the top, and they are not doing anything significant there. The suggestion originally made by Sigworth was that the high resolution of the original measurements by Conti & Stühmer didn't actually prove that there was some sort of time correlation. One thing this graph says rather clearly is that there is cooperativity between the four domains; it is this last cooperative event that gives the 2.3 charges.

Horn: I don't agree with that. I don't believe that the cooperativity is as dramatic or as large as he thinks. All the other evidence says that cooperative interaction between the various voltage sensors in the different domains is relatively small. There may be some cooperativity, but Isacoff's estimate is much larger than anyone else's, and certainly larger than our estimate.

Strichartz: Your traces from squid implied a rising phase to gating currents. Can you really resolve that, given the limitations of the rate at which you can truly change the membrane potential? And if there really is a rising phase, implying that gating currents move or develop more slowly at first, and gating particles have some inertia involved with them, what does this imply about the distribution of kinetic steps?

Keynes: We have got a lot of data, and we spent a long time trying to fit the standard multistate model that Zagotta et al (1994) used for four S4 voltage sensors making transitions in parallel between three states. It is necessary to introduce some cooperativity between the first two steps to get that kinetic to come out. It then comes out without an excessive level of cooperativity, and you can fit the data quite well. The only trouble with fitting these kinds of data is that the model we were working with had 36 states. If you are trying to do any standard fitting, therefore, you have multiple minima to play with and it is not possible to arrive at unique fits. We did finally fit the data quite nicely in the paper by Keynes & Elinder (1998a). There was a great deal of argument earlier about the currents in the slowly rising phase, because Stimers et al (1987) had claimed that it was an artefact due to incomplete compensation for the Schwann cell series resistance in setting up the voltage clamp. However, this certainly did not apply to the observations of Keynes et al (1990), made with very careful adjustment of the feedback, or to the results of Keynes & Elinder (1998a). There had also been a severe quarrel between Bezanilla and Armstrong, because Armstrong & Gilly (1979) published a paper showing that there was a slow rising phase, and Bezanilla said it wasn't there. In fact, Gilly and Armstrong didn't have nearly as good resolution in their recordings as we have now, but what we found was exactly what they said. I think the slow rising phase is a reality, and the resolution of our recordings is quite good enough to show this.

Horn: It is clear that there is a rising phase in Shaker mouse K^+ channels. There, the rising phase is much slower, and you can do the recordings in patches that have much higher frequency resolution than you can get from squid axon or cut open oocytes. There is no question that there is a slow rising phase.

Keynes: When I was last looking through the experimental data I couldn't locate enough K^+ channel records to be very convinced.

Horn: It's absolutely convincing, and Bezanilla's data are some of the best to show this. If he didn't believe it before, I'm sure he believes it now.

Segal: One of the most impressive things about the prolonged bursts of the Na^+ channel is that very frequently the bursts will start up again with the next pulse. I don't know how to interpret this. Do you understand this as a particular closed kinetic state that predisposes the channel to start up again in a prolonged burst, or is there any evidence for covalent modification such as interaction with actin that would explain some of the persistence of bursting after coming back seconds later and giving another voltage pulse?

Keynes: I don't know the answer to that from the Patlak & Ortiz data. In squid, in the models that we have built, we can explain everything kinetically and predict everything by assuming that there is a voltage-dependent inactivation that takes place, but you can get out of it to a further open state.

Horn: One of the ways that Patlak describes this is that the inactivation gate gets stuck for some period of time. His analysis says that there is no inactivation for the channel: it will activate with the normal activation kinetics and voltage dependence, but the inactivation gate won't work at all. If you keep repeating pulses over and over again, that same single channel will not be inactivated at all, other than perhaps by slow inactivation. Then at some point the channel recovers and functions normally. It is as if the gate is somehow hung up for long periods— many seconds.

Segal: The current from these prolonged bursts of the Na^+ channel is truly impressive, and in pathological states it may turn out to be important to understand.

Cummins: An important factor in understanding inactivation and the generation of persistent currents may be subtle differences between the skeletal muscle and neuronal channels or isoforms. Patlak & Ortiz (1986) found that non-inactivating or persistent Na^+ currents in skeletal muscle cells were very small, whereas in the squid axons persistent Na^+ currents are much larger. There seem to be subtle but important differences between the different isoforms. There may be differences in cooperativity. All of this has to be worked out very carefully.

Horn: Is there any disease mutant that shows that characteristic—a persistent current like that?

Segal: There is a prolonged bursting mode in one form of the cardiac long-QT syndrome, LQT3, in which there is a mutation in the SCN5A Na^+ channel α subunit (Bennett et al 1995).

Horn: I wonder whether the hyperkalaemic periodic paralysis is an example of that.

Bean: I think that does behave this way. There are a lot of late openings.

Horn: It is just as if during the long depolarization the inactivation is not there.

Bean: I am not sure that it has been analysed to that level, whether it is incomplete or whether the inactivation of a single channel is not there.

Cummins: I would argue against truly non-inactivating channels playing a role in hyperkalaemic periodic paralysis. Early studies indicated that this was the case, but no one has really shown this with the hyperkalemic mutations.

Waxman: Ted Cummins, you have thought a lot about the slowness of closed-state inactivation in the hNE channel, and Bruce Bean, you have thought about the resurgent current produced putatively by $Na_v1.6$. Are there any clues in either of those sets of observations?

Cummins: The striking feature to me of that closed state inactivation is the difference between, say, the PN1 neuronal channel and the skeletal muscle. It is a major difference, but I don't know what the underlying mechanism is.

Bean: We think of the resurgent current as being connected with a second inactivated state that recovers through the open state. The open state would therefore be the same as the usual open state, which would be in contrast to the model for the squid axon, where the idea is that there is a second open state kinetically.

I have a question about the diversity of channels, and the degree to which one sees incomplete inactivation in different Na^+ channels. As Richard Keynes has pointed out, this is seen in many channels other than the squid channel. The other striking characteristic about the squid channel is that you get more occupation of the late open state with increasing depolarization. I would ask the group as a whole: are there other cases where this increase in late openness is seen as you depolarize beyond about −10 mV? Have people looked for this with the cloned channels, to see whether the incomplete opening increases with depolarization? It seems to me to be a distinctive property of the squid axon sodium channel. It is not obvious in the cases of mammalian neurons that I have looked at, even though we have looked for this occasionally in hippocampal neurons.

Horn: It is not an obvious feature of the human skeletal muscle Na^+ channel, which is the one that we primarily use.

Strichartz: It is something that is seen in frog node. With normal Na^+ gradients this is something that is seen as channels are driven through the reversal potential to yield outward current. I wondered whether this might not be due to the possibility that the last closed state—or the pre-open state, if you will—can go to an 'inactivated' state, a pathway which may be relatively favoured for the smaller depolarizations, but larger depolarizations may inactivate channels faster from the open state, an inactivation state that may be more reversible.

Goldin: At the macroscopic level we do see the differences you describe for different isoforms. Type 6 ($Na_v1.6$) shows a greater percentage of persistent or sustained current at more positive potentials, whereas with type 1 ($Na_v1.1$) we see a greater percentage of persistent current at more negative potentials.

Bean: As Gary Strichartz has pointed out, it might be worthwhile to look at channel behaviour as the voltage changes systematically through the reversal potential, to see whether the behaviour changes with outward current. It easily could, if there were interactions between the permeating ions and the inactivation machinery.

Keynes: We have a great advantage in squid axons in being able to treat them more roughly than is possible with patches in single channels. This meant that I was able to show that the equilibrium potential is 95 mV more positive than the one for the initial opening, which is a huge difference.

Bean: That means that the ionic selectivity is actually different for the different open states.

Keynes: Not necessarily. There was no way we could establish this.

Horn: If the selectivity isn't changed, it must be a leakage current. It seems these are the only possibilities, if you have a shift of an experimental reversal potential of 95 mV: either the selectivity is changing or some other conductance has appeared in parallel.

Keynes: I would refer you to the model that Elinder and I produced in 1998 (Keynes & Elinder 1998b). This is quite a simple scheme that fits the data nicely.

Bostock: I have a possibly naive general question. In Chandler & Meves' (1970a,b) original experiments, referred to above, they found that in axons perfused with fluoride ions, part of the Na^+ current failed to inactivate, or inactivated only very slowly, whereas this had not previously been observed in intact axons. Fluoride also irreversibly knocked out most of the K^+ currents. We have all been thinking about the Na^+ channel on its own, but are anions important? Is it now understood what the role of fluoride was in those experiments?

Horn: We also use fluoride, and we don't see that late outward current. For the particular channels we use, under our conditions, we don't see the late current.

Keynes: We can't get axons to survive without having fluoride present.

Horn: That's the main reason we use fluoride: the cells last longer.

Baker: If I can get this straight, there are two ideas. Either the Na^+ channels go through a fast inactivated state in order to give rise to a late or persistent current, or the channels simply lose fast inactivation. One thing that Hugh Bostock and I noticed when we were recording from sensory neurons was that the late current comes in at more negative potentials than the transient current. I noticed also that Professor Crill and others have seen the late current coming in at more negative potentials than the transient current. One explanation is that the removal of fast inactivation reveals a low threshold current. This is predicted by the Armstrong–Bezanilla–Aldrich–Corey–Stevens model of the Na^+ channel (e.g. Gonoi & Hille 1987). But if one compares professor Keynes' model, it is the voltage dependence of inactivation (h) that will give rise to the low threshold nature of the late current, because by entering an open state by way of the fast inactivated state, the late current activation might be expected to follow the voltage-dependence of h. In sensory neurons, if one records h for the transient current, it turns out that the voltage dependence of activation of the late current is very similar to the voltage dependence of inactivation of the transient current.

Cummins: At least from my evidence concerning closed state inactivation of Na^+ channels (Cummins et al 1998), it might be that the inactivation is dependent on the channels opening for the peripheral neuronal channels, as opposed to skeletal muscle channels, which are more likely to inactivate from the closed states. This may be why you see such a concordance between the Hinf (steady-state inactivation curve) and the voltage-dependence of the persistent current: the peripheral neuronal Na^+ channels may have to open if they are going to inactivate.

References

Armstrong CM, Gilly WF 1979 Fast and slow steps in the activation of sodium channels. J Gen Physiol 74:691–711

Bennett PB, Yazawa K, Makita N, George AL Jr 1995 Molecular mechanism for an inherited cardiac arrhythmia. Nature 376:683–685
Chandler WK, Meves H 1970a Sodium and potassium currents in squid axons perfused with fluoride solutions. J Physiol 211:623–652
Chandler WK, Meves H 1970b Evidence for two types of sodium conductance in axons perfused with sodium fluoride solution. J Physiol 211:653–678
Cummins TR, Howe JR, Waxman SG 1998 Slow closed-state inactivation: a novel mechanism underlying ramp currents in cells expressing hNE/PN1 sodium channels. J Neurosci 18: 9607–9619
Gonoi T, Hille 1987 Gating of Na^+ channels. Inactivation modifiers discriminate among models. J Gen Physiol 89:253–274
Keynes RD, Elinder F 1998a On the slowly rising phase of the sodium gating current in the squid axon. Proc R Soc Lond B Biol Sci 265:255–262
Keynes RD, Elinder F 1998b Modelling the activation, opening, inactivation and reopening of the voltage-gated sodium channel. Proc R Soc Lond B Biol Sci 265:263–270
Keynes RD, Greeff NG, Forster IC 1990 Kinetic analysis of sodium gating current in the squid giant axon. Proc R Soc Lond B Biol Sci 240:411–423
Patlak JB, Ortiz M 1986 Two modes of gating during late Na^+ channel currents in frog sartorius muscle. J Gen Physiol 87:305–326
Stimers JR, Bezanilla F, Taylor RE 1987 Sodium channel gating currents. Origin of the rising phase. J Gen Physiol 89:5621–5640
Zagotta WN, Hoshi T, Aldrich W 1994 *Shaker* potassium channel gating. III: evaluation of kinetic models for activation. J Gen Physiol 103:321–362

Molecular basis for function in sodium channels

Richard Horn

Department of Physiology, Institute of Hyperexcitability, Jefferson Medical College, 1020 Locust Street, Philadelphia, PA 19107, USA

Abstract. Na^+ channels earned their unique role in excitable cells because of two functional properties, finely honed by evolution. The first is their exquisite sensitivity to small changes of membrane potential: a depolarization of only 10 mV can increase open probability by as much as two orders of magnitude. The second is the rapidity with which they respond to changes of membrane potential: their gates begin to open tens of microseconds after a depolarization. These features are built into two sets of moving parts: voltage sensors that respond directly to changes of membrane potential, and gates that open and close in response to voltage sensor movement. We have explored these movements using a combination of electrophysiology, site-directed mutagenesis, cysteine accessibility scanning and photoactivated cross-linking using a bifunctional cysteine reagent. The main voltage sensors of Na^+ channels are four homologous S4 segments, each of which has a unique functional role. These transmembrane segments are almost completely surrounded by hydrophilic crevices. The membrane electric field moves these positively charged helices through a short, hydrophobic 'gating pore'. The minimum contact between an S4 segment and its gating pore insure that a small movement can rapidly move several of its charged residues across the electric field.

2002 Sodium channels and neuronal hyperexcitability. Wiley, Chichester (Novartis Foundation Symposium 241) p 21–33

Sodium channels earned their unique role in excitable cells because of two functional properties, finely honed by evolution. The first is their exquisite sensitivity to small changes of membrane potential: a depolarization of only 10 mV can increase open probability by as much as two orders of magnitude (Hirschberg et al 1995). The second is the rapidity with which they respond to changes of membrane potential; their gates begin to open tens of microseconds after a depolarization (Hodgkin & Huxley 1952). These features are built into two sets of moving parts, voltage sensors that respond directly to changes of membrane potential, and gates that open and close in response to voltage sensor movement (Yellen 1998). It is generally believed that voltage sensors and gates are separate entities in the major superfamily of voltage-dependent ion channels, including Na^+, Ca^{2+} and K^+ channels. Moreover, there are excellent candidates for each type of moving part in specific regions of the channel protein, as described below. My laboratory has explored these movements using a combination of electrophysiology, site-directed mutagenesis, cysteine accessibility scanning, and photoactivated cross-linking using a bifunctional cysteine reagent. Some of our results are summarized below.

Voltage-gated ion channels have an approximately fourfold radial symmetry with each domain, or subunit, containing six putative transmembrane segments, S1–S6. K^+ channels are made of four subunits arranged around a central permeation pathway. In Na^+ and Ca^{2+} channels four homologous domains of a single polypeptide are arranged around the permeation pathway. The ion-selective permeation pathway is lined primarily by the four S6 segments and the extracellular S5–S6 loops. The main voltage sensors are the four positively charged S4 segments, each of which has four to eight basic residues, either arginines or lysines, which are usually separated from each other by two neutral residues. Depolarization is expected to move S4 segments outward through the electric field (Keynes 1994, Sigworth 1994, Yellen 1998, Keynes & Elinder 1999, Bezanilla 2000). One early consequence of this S4 movement is the opening of the activation gate, believed to be formed by the cytoplasmic ends of the channel's four S6 segments, at the entrance of the permeation pathway (Liu et al 1997, Del Camino et al 2000). Prolonged depolarization also causes the inactivation gates, located elsewhere in the protein, to close. Each S4 segment of a Na^+ channel has a unique functional role, as shown by systematic mutagenesis (Chen et al 1996, Kontis & Goldin 1997, Kontis et al 1997). The S4 segments of domains 1 and 2 (D1/S4 and D2/S4) primarily underlie the process of activation. D3/S4 and D4/S4 also contribute to the process of fast inactivation.

Cysteine scanning of S4 movement

We initially examined the movement of D4/S4 of the human skeletal muscle Na^+ channel by cysteine accessibility scanning (Yang & Horn 1995). This study was motivated by the fact that a point mutation (R1448C) linked to the disease paramyotonia congenita replaces the outermost charged residue of D4/S4 with a cysteine (Ptacek et al 1992). Exposure of this cysteine mutant to the hydrophilic cysteine reagent methanethiosulfonate-ethyltrimethylammonium (MTSET) causes a pronounced alteration of inactivation kinetics (Yang & Horn 1995). The accessibility of this cysteine to extracellular MTSET is voltage dependent. Depolarization causes the exposure of the introduced cysteine on the extracellular surface of the Na^+ channel, and hyperpolarization buries this residue. The steepness of the voltage dependence and the kinetics of exposure are consistent with the idea that D4/S4 is one of the voltage sensors underlying gating.

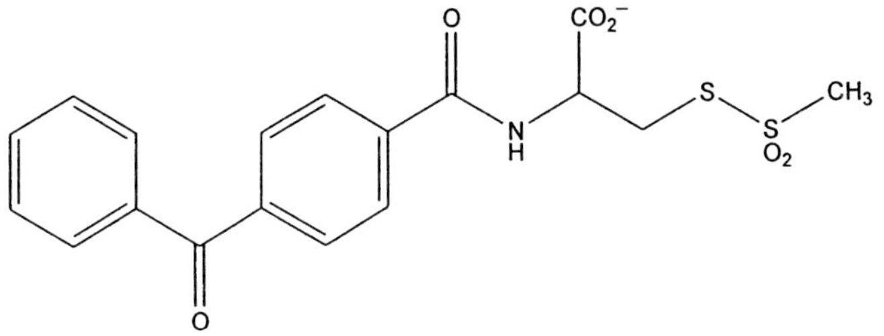

FIG. 1. The structure of benzophenone-4-carboxamidocysteine methanethiosulfonate (BPMTS).

We have substituted cysteines along the full length of D4/S4 and were surprised to find that, although D4/S4 is a true transmembrane segment, it is almost completely surrounded by hydrophilic crevices (Yang et al 1996, 1997, Mitrovic et al 2000). The membrane electric field moves these positively charged helices through a short, hydrophobic 'gating pore'. The minimum contact between an S4 segment and its gating pore insures that a small movement can rapidly translocate several of its charged residues across the electric field.

Immobilizing moving parts

Recently we have been exploring the possibility of systematically immobilizing the moving parts of voltage-gated ion channels and examining the biophysical consequences on their function. To this end we designed a bifunctional cysteine reagent, benzophenone-4-carboxamidocysteine methanethiosulfonate (BPMTS), that will attach covalently to introduced cysteines.

While recording currents from these labelled ion channels, we expose them to ultraviolet (UV) light, which causes the ketone group on the benzophenone to insert covalently into neighbouring C–H bonds, with the goal of preventing the relative movement between two secondary structural regions of the channel.

Before testing this method on an S4 segment, we did a proof of-principle experiment by immobilizing a gate. We chose the fast inactivation gate of the Na^+ channel, believed to be the cytoplasmic linker between domains 3 and 4 (Catterall 2000). We substituted a cysteine for an isoleucine in the critical triad isoleucine-phenylalanine-methionine (IFM) within this linker. We then reacted this IFM/CFM mutant with BPMTS. UV light could selectively immobilize the labelled inactivation gate open at a hyperpolarized voltage and shut at a depolarized voltage, as predicted (Horn et al 2000). This supports the feasibility of using this technique on another moving part, a voltage sensor.

To demonstrate that we could immobilize a voltage sensor, we first turned to Shaker potassium channels, because (i) gating currents, which primarily reflect S4 movement, are easy to measure, and (ii) the channels have tetrameric symmetry. We used a non-inactivating variant of Shaker into which we introduced the cysteine mutant A359C, located at the extracellular end of the S4 segment. After exposure to BPMTS each channel has four labelled S4 segments.

We wanted to test if photo cross-linking would completely immobilize an S4 segment. We expected that the reduction of ionic current in response to UV would be more rapid than reduction of gating current through the same channels, because immobilizing a single S4 segment should prevent a channel from opening, whereas it would only reduce a fraction of the channel's gating current, assuming that other S4 segments remained capable of moving.

We found that the reduction of gating current in response to UV irradiation was exponential with time, and that the corresponding reduction of ionic current measured in the same cells followed the fourth power of that exponential function (Horn et al 2000). This result suggests that (i) photo cross-linking a single S4 segment completely prevents its ability to move charge, and (ii) immobilizing one S4 segment has no effect on the charge movement of S4 segments in other subunits, i.e. S4 segments move independently of one another.

Our data with Shaker and with the inactivation gate support the idea that immobilization is complete (the moving part is not just impeded from moving; movement is stopped cold) and has a local effect (immobilization of one moving part doesn't prevent other movement within the same channel).

Immobilization of Na^+ channel S4 segments

Armed with the aforementioned control experiments we examined the consequences of immobilizing individual S4 segments of Na^+ channels (Horn et al 2000). We chose D2/S4 and D4/S4 for the first experiments, in both cases using cysteine mutants of basic residues.

Photo cross-linking of D2/S4, using the cysteine mutant of the outermost arginine, had a similar effect as immobilization of the Shaker S4 segment, namely it reduced the number of functional channels with increasing exposure to UV. This is the expected consequence of immobilizing a voltage sensor that is coupled to activation gating. By contrast photo cross-linking D4/S4 produced effects on both activation and inactivation. The effects depended on the specific residue labelled with BPMTS and on where BPMTS was applied, extracellularly or intracellularly. If cysteines were labelled with extracellular BPMTS, the consequence of UV irradiation was primarily a slowing of inactivation. However if the same residue were labelled with intracellular BPMTS, irradiation reduced the current.

We interpret these results as follows. D4/S4 makes two sequential movements in response to a depolarization. The first is coupled to an activation gate, the second to the inactivation gate. Immobilization of D4/S4 from the inside prevents the first step, and therefore prevents channels from opening. Immobilization of D4/S4 from the outside only prevents the second step, which retards the closing of the inactivation gate. This interpretation provides insight into the efficiency of Na^+ channel gating. The delay in the onset of inactivation after a depolarization, due to the sequential movement of D4/S4, allows activation gates to open before the inactivation gate closes.

Conclusion

The last great molecular puzzle about Na^+ channel gating is how the moving parts talk to one another. How does translocation of S4 segments through their gating pores cause gates to open and close? There are two competing hypotheses for coupling mechanisms between these moving parts (Horn 2000). The first is that S4 movement pulls or twists on the linkers that connect it with the S3 and S5 segments. The second is that S4 movement causes a compensatory repacking of the transmembrane segments that surround it. We expect that this puzzle will be cracked in just a few years as a variety of molecular and biophysical tools are brought to bear on the problem.

References

Bezanilla F 2000 The voltage sensor in voltage dependent ion channels. Physiol Rev 80:555–592
Catterall WA 2000 From ionic currents to molecular mechanisms: the structure and function of voltage-gated sodium channels. Neuron 26:13–25
Chen LQ, Santarelli V, Horn R, Kallen RG 1996 A unique role for the S4 segment of domain 4 in the inactivation of sodium channels. J Gen Physiol 108:549–556
del Camino D, Holmgren M, Liu Y, Yellen G 2000 Blocker protection in the pore of a voltage-gated K^+ channel and its structural implications. Nature 403:321–325
Hirschberg B, Rovner A, Lieberman M, Patlak J 1995 Transfer of twelve charges is needed to open skeletal muscle Na^+ channels. J Gen Physiol 106:1053–1068
Hodgkin AL, Huxley AF 1952 Currents carried by sodium and potassium ions through the membrane of the giant axon of *Loligo*. J Physiol 116:176–195
Horn R 2000 A new twist in the saga of charge movement in voltage-dependent ion channels. Neuron 25:511–514
Horn R, Ding S, Gruber HJ 2000 Immobilizing the moving parts of voltage-gated ion channels. J Gen Physiol 116:461–475
Keynes RD 1994 The kinetics of voltage-gated ion channels. Q Rev Biophys 27:339–434
Keynes RD, Elinder F 1999 The screw-helical voltage gating of ion channels. Proc R Soc Lond B Biol Sci 266:843–852
Kontis KJ, Goldin AL 1997 Sodium channel inactivation is altered by substitution of voltage sensor positive charges. J Gen Physiol 110:403–413 (erratum: 1997 J Gen Physiol 110:763)

Kontis KJ, Rounaghi A, Goldin AL 1997 Sodium channel activation gating is affected by substitutions of voltage sensor positive charges in all four domains. J Gen Physiol 110:391–401
Liu Y, Holmgren M, Jurman ME, Yellen G 1997 Gated access to the pore of a voltage-dependent K^+ channel. Neuron 19:175–184
Mitrovic N, George AL Jr, Horn R 2000 Role of domain 4 in sodium channel slow inactivation. J Gen Physiol 115:707–718
Ptacek LJ, George AL Jr, Barchi RL et al 1992 Mutations in an S4 segment of the adult skeletal muscle sodium channel cause paramyotonia congenita. Neuron 8:891–897
Sigworth FJ 1994 Voltage gating of ion channels. Q Rev Biophys 27:1–40
Yang N, Horn R 1995 Evidence for voltage-dependent S4 movement in sodium channels. Neuron 15:213–218
Yang N, George AL Jr, Horn R 1996 Molecular basis of charge movement in voltage-gated sodium channels. Neuron 16:113–122
Yang N, George AL Jr, Horn R 1997 Probing the outer vestibule of a sodium channel voltage sensor. Biophys J 73:2260–2268
Yellen G 1998 The moving parts of voltage-gated ion channels. Q Rev Biophys 31:239–295

DISCUSSION

Catterall: Your experiments seemed to imply that even when D4/S4 segments are cross-linked, the channels will still inactivate, because you get slow inactivation. It isn't that you prevent a fraction of the channels from inactivating; rather they all inactivate slowly. How do you envisage this? It means that the S4 segments must be fixed in space, and yet inactivation can still occur.

Horn: Inactivation does still occur, but it is inefficient.

Catterall: Is it inefficient in terms of voltage, or is it just slow?

Horn: It is just slow.

Catterall: So it's not shifted.

Horn: The steady state is shifted, but I would say that in order for the inactivation to occur efficiently or in the right voltage range, you need the full range of movement of that S4 segment. Whether that is due to the inactivation gate binding inefficiently without that S4 movement, I don't know.

Catterall: It means that it is not a hard and fast coupling mechanism.

Horn: That's right. On the other hand, the initial movement looks like it is a hard and fast coupling. If you stop that S4 segment from moving either in Shaker or domain 2 or domain 4 of the Na^+ channel, the channels fail to open on the activation site.

Wood: Is it possible to identify the residues that the benzophenone actually cross-links?

Horn: That is what we are doing now. It is not an easy experiment. We are not doing it with Na^+ channels because we can't generate enough of them, so we are using K^+ channels. We are transfecting lots of dishes, labelling lots of channels and we are able to get a large amount of protein — in the order of 10 picomoles. We then irradiate these dishes, do a trypsin digest and send them off to the mass spectrometer, to find out where the benzophenone is inserting. It would be wonderful to have this information because at different membrane potentials this is presumably linking at different places, and we ought to be able to find out how the S4 segment is oriented with respect to the other transmembrane segments around it. We are quite excited about what the biochemistry is going to teach us.

Bean: In the D4/S4–R3, when you cross-link this and it is on the outside, the current gets bigger. Do you also see a change in the voltage dependence of activation? Intuitively it seems that if you lock that activation gate open, you should see a change in voltage dependence.

Horn: We do see a change in the voltage dependence of activation, but I think it can be explained without doing extensive kinetic modelling, which I don't trust too much in these experiments. What you see is that they open at more negative voltages. I think this can be explained entirely by inhibiting inactivation. Ordinarily, these channels will inactivate from closed states to some extent. If you inhibit that, it will tend to shift the apparent activation to more negative voltages. Another reason we haven't done extensive kinetic modelling is that we know that the cross-linking isn't 100% efficient. We have a heterogeneous population after we give UV irradiation. Also, the labelling itself affects the kinetics.

Strichartz: When you did the cross-linking in the condition where the inactivation gate was in the closed state, and you showed that there was a reduction in peak current, it seemed to me that the sustained current for that particular trace was unaffected by that procedure. Did you notice that?

Horn: Yes, the sustained current actually gets bigger. It looks as if the inactivation gates of some of the channels have been locked open. We also see a reduction of current when we try to irradiate at hyperpolarized voltages. This may be due to some inactivation gates being locked shut. It is not 100% efficient. First of all, the gates are not 100% determinant. At a hyperpolarized voltage, not all of the inactivation gates are open, and at a depolarized voltage not all are shut. The effect we see, however, is not as clean as you would expect from the biophysics.

Strichartz: When you apply the illumination for photocross-linking, during the few milliseconds of the depolarized pre-pulse is it applied repetitively?

Horn: We have done it several different ways, but for those experiments we gave a 2 s depolarization to 0 mV, waited 100 ms and then applied UV light. During this entire time, the inactivation gates were primarily closed, although as you saw there is still a little bit of steady-state current. Some of the gates are open and they could get cross-linked open.

Strichartz: When you say 'during this 2 s the inactivation gates are primarily closed', does this mean that over that period every channel has some probability that its inactivation gate is open? Or is there some subpopulation that doesn't have the opportunity to close during that period and therefore isn't photoreactivatable in the closed state?

Horn: There is a probability that the inactivation gates are open during those long depolarizations. You can see there is a small steady-state current, although for 2 s that current decays pretty much to zero. There is a slow inactivation gate, which is also involved in that. We have done this experiment in many different ways. Specifically, we have given briefer exposures to UV light repetitively to avoid the problem of the slow inactivation. We get similar results as you saw here.

Strichartz: Can you give a brief enough pulse, let's say, during the end of the depolarization when the channels are in this steady state and you have sustained current only, to see whether you get some selective state-dependent labelling?

Horn: That is a little complicated for a reason that I didn't mention. This is because the cross-linking doesn't happen as fast as one might think. It happens slowly. We didn't know this until we started using UV flash lamps instead of continuous exposure to UV light. For example, in the Shaker K^+ channel, if you give a flash of UV light, the time constant for the cross-linking reaction is about 200 ms.

Strichartz: Is that the lifetime of the photoactive intermediate?

Horn: An early intermediate is a triplet biradical state. Then there are two steps after this: first, it pulls off a hydrogen from a —CH bond, and then there are two carbon free radicals (a ketyl and an α free radical) which have to recombine. This recombination step is faster in Na^+ channels, in the order of 20 ms.

Keynes: In the model that Elinder and I produced, the effect of inactivation was to stop the hydration of the channel. Is this ruled out?

Horn: Absolutely not.

Keynes: Your reagent could be blocking this from happening.

Horn: I would say that the hydration is way upstream of the effect we are having. We are actually preventing the S4 segment from moving. This means that anything that follows S4 movement will be stopped, including a hydration near the activation gates.

Keynes: You have shown that the total transfer of charge is reduced when you block.

Horn: That is something we showed in Shaker. We showed that the immobilization of a single S4 segment almost completely immobilizes its charge. This was actually a little surprising for us: this reagent is not infinitely small: between the cysteine and the ketone of the benzophenone is about 10.4 Å, and so the movement has to be large enough so that if there is a cross-link, you will see an effect. Apparently it is large enough. You might think that there is some flexibility in the linker, but this doesn't seem to prevent this complete immobilization of the S4 movement.

Spruston: In the Shaker experiments that you showed, there didn't seem to be a change in the kinetics of the gating current. Have you tried immobilizing multiple S4 domains to see if the kinetics change, suggesting possible interaction between the movement of the S4 domains?

Horn: They are all labelled in that experiment. You are right; this is exactly what we were hoping for. What this shows is that we are not just slowing down the movement: we are stopping it cold. What you are looking at are the other S4 segments that are not immobilized. In those experiments, we can estimate the number of S4 segments that immobilize per channel. About 30% of them have multiply immobilized S4 segments. We are glad to see that the kinetics didn't change. This means that if one S4 segment is immobilized, its neighbours continue to carry gating charge with the same kinetics.

Spruston: Doesn't that imply that the S4 segments are moving independently?

Horn: Yes.

Goldin: Do you believe there is no cooperativity, or is that coming later?

Horn: We can estimate from our experiments the amount of cooperativity. This is an unusual case: we are asking how much gating charge is immobilized in another S4 segment if you immobilize one of them. Our estimates are in the order of 4–5%, which is small. Does that mean that when an S4 segment moves in a normal channel which is not labelled and which doesn't have cysteine on it, that there is 5% cooperativity? We can't say for sure. All I can say is that if I immobilize one completely, it doesn't seem to affect the ability of the others to carry their total charge. There are many different ways of measuring cooperativity, and the results don't agree with one another in terms of the degree that is present. Isacoff's measurements have the largest amount: about half of the total charge movement is cooperative here. But all the other studies suggest it is much less than that.

Goldin: You hope to identify a specific residue. Do you really think it will be a specific residue? You have 200 ms; it could be binding to multiple residues.

Horn: The 200 ms is after the photoreaction has started. There is an initial photoreaction where you extract a hydrogen, which takes place within 2 ms. After this, the only things that can cross-link are the benzophenone and the particular residue from which it extracted its hydrogen.

Goldin: So the target is chosen within 2 ms?

Horn: Yes, but that doesn't mean that the benzophenone isn't sitting in a bunch of other places. In biochemical studies that have used this reagent, it tends to label in a specific place. We are fairly hopeful that we are going to find a single residue under particular conditions — for a given voltage, for example.

Goldin: There was one thing you didn't mention: what about coupling between the voltage sensor and the inactivation gate?

Horn: That was the initial idea. The point is, we can begin to use this reagent in different places that potentially could be important for coupling. All we have done so far is to show that it is possible to immobilize the gate and the voltage sensor. There are many other putative sites that could be important for coupling, for example linkers. There are other things that we can look at that have less to do with coupling, but more to do with where the gates are under different conformations. One of the things that surprised us is that we could immobilize the inactivation gate. This means that the inactivation gate is not just hanging in the breeze when it is open, but is moving from one protein surface to another protein surface. We can cross-link it to either one of those. Another possibility is to try to locate where the inactivation gate is when it is open. This may turn out to be important for some of the things that we talked about earlier, which is cases where the inactivation gate is not able to close very efficiently, for example in a disease state, or in certain isoforms of Na^+ channels. It may be because of an interaction between the inactivation gate and some other proteinaceous region when it is open.

Strichartz: Have you tried preventing that reaction with quenchers of the photoactivated state that would be soluble in the intracellular solution? If the gate is in equilibrium between the unassociated form and a bound form, in principle you could quench that reaction by trapping it in the free state.

Horn: We haven't tried using quenchers. Most of the time we are trying to get as high an efficiency of cross-linking as possible. We have thought about ways of making the efficiency higher. There are competing reactions that could be affecting the efficiency. One of them is that we make carbon free radicals in this process, and they could be oxidized by molecular oxygen. We thought about argon-loading our solutions to increase the efficiency, but we haven't done this yet.

Bean: What is your mental picture for what is happening when the photoreaction occurs with the inactivation gate in the closed state? How does the label change inactivation?

Horn: I don't know how it extends from this particular residue. The best structural information is from Bill Catterall's laboratory, that has published the NMR structure of the 3–4 linker. When we started doing these experiments, initially we didn't label that particular cysteine: we started out with the phenylalanine. In that case, when we labelled with the photoreagent, the labelling itself almost completely abolished the inactivation. It was an unusable mutant. We ended up labelling the isoleucine instead, changing that to cysteine. This left quite a bit of inactivation. From their studies, the phenylalanine and the cysteine are pointing in almost the same direction, in a hydrophilic part of the inactivation gate. They are not pointing in exactly the same direction, but they are extending out from one face. It is labelled already, before we do the experiment, so when the inactivation gate is closing it is dragging the benzophenone. The benzophenone part itself is relatively hydrophobic, so it probably finds some nice hydrophobic place to lie down on, and this reagent usually cross-links to the peptide backbone.

Keynes: Does your blocking stop the possibility of re-opening?

Horn: If you cross-link it shut, it prevents it from opening, and locks the inactivation gate permanently shut.

Cummins: Is there anything distinctive about the sequence of that linker that might make it the target for regulation by some intrinsic molecule?

Horn: It is on the inside, so it would have to be some cytoplasmic molecule.

Catterall: There is a protein kinase C site that is phosphorylated and has subtle effects. We have looked for proteins that interact with the 3–4 linker, but we haven't found any in a range of protein–protein interaction assays. We have considered the idea that something inside the cell binds the gate in an open conformation and keeps it open. This may be true, but we haven't been able to demonstrate it.

Cummins: Coming back to the domain 4 mutations, in the disease mutant R1441C (which is the same as your R1C mutant), Featherstone et al (1998) proposed that deactivation is altered by this mutation, and that the defective deactivation is a major contributor to the hyperexcitability in paramyotonia congenita. Did you look at deactivation with the domain 4 mutants?

Horn: Yes, it is affected. I don't know how to evaluate how important this is, but we saw a similar effect on deactivation as they did.

Cummins: When you put the label on, could you see a change in deactivation?

Horn: We didn't measure deactivation when it was labelled. One thing about the deactivation is that it is another piece of evidence that this S4 segment plays some role in the activation process. There is some contention about that, especially from Dottie Hanck: she believes that this is purely a voltage sensor for inactivation, and that it doesn't play any role in the activation process. I don't think that is possible. The other three S4 segments can't play the entire role for activation. Partly, this is because in the K^+ channels, all four of them are playing an important role for activation: it would be surprising if it were fundamentally different in the Na^+ channel. Between the data on deactivation and the data we have shown with cross-linking, it is clear that D4/S4 is doing something for the activation process as well as the inactivation process.

Wood: Can you use your cysteine-modifying reagents to map the sites of interaction of accessory subunits?

Horn: It is not easy. With the cysteine reagents, if there is an effect of labelling there are biophysical consequences. This just tells us that the residue is accessible on one side or another: it doesn't tell us that it is involved in an interaction. If you wanted to find an interaction site with cysteines, you might have to try to do cysteine cross-linking (i.e. disulfide formation), and this is almost hopeless. It is a real fishing expedition. They have to get reasonably close to cross-link. Some people have recently been using bifunctional cysteine reagents to see if they can find cross-linking. Even in that case, if you were able to cross-link an α subunit with a β subunit, you wouldn't know it unless there were an effect. The chances of there being an effect are pretty small.

Isom: I will show in my paper (Isom 2002, this volume) that if you knock out $\beta 1$ or $\beta 2$ you do see functional effects.

Horn: Knocking it out is different from whether you covalently link it.

Isom: It depends on which step they are actually interfering with because they are not co-translationally assembled and linked. We think it is one of the last steps before insertion into the membrane. I am not sure how you are going to be able to interfere with that before the coupling happens.

Strichartz: I wonder if the disparity among the different reports about the degree of coupling and interaction between subunits might not have some relationship to differences in initial conditions. I am reminded of the Cole–Moore effect, which was first seen in squid axons and has appeared elsewhere. This was an observation that if you greatly hyperpolarize the squid axon and follow the delayed rectifier currents, you can fit its kinetics by n^{23} (an exponential rise raised to the 23rd power), rather than the n^4 or n^6 seen from a normal resting potential. The suggestion is that the way the different K^+ channel subunits interact depends on where they are coming from to begin with. Perhaps if extreme initial conditions lined up subunits in the exact same initial orientation (and we don't know whether squid K^+ channels are homotetramers or not) we might be able to discriminate the dependence on the interactions between them. But if they are in relatively different conditions, the different subunits within a particular channel may have less of a probability of interacting with each other during a single gating transition.

Horn: That is possible. When you do an experiment like the Cole–Moore experiment you are moving way beyond the physiological range. It might be that under normal physiological conditions, because the initial conditions are not so severe, we might see either more or less cooperativeity.

Goldin: How consistent is your gating with Bezanilla's crevice?

Horn: Our data completely disagree with one aspect of the Bezanilla model: our results indicate that the movement cannot be a pure rotation. There are crevices or vestibules in all the models. Richard Keynes' model has large vestibules, as does Isacoff's model and also ours. It is the orientation of those vestibules that will determine whether you can get rotation to transfer charge, or translation to transfer charge. In order to get pure rotation of the S4 segment to transfer charge, you have to change the orientation of the electric field. The electric field is ordinarily in the plane of the membrane. To get a rotation to do this, you have to move it perpendicularly to the plane of the membrane. Then you have rotated from one side of the α helix to the other. The electric field then moves through the S4 segment, but the S4 segment is part of the electric field, and you are moving it from one side to the other. Then you need to have vestibules lined up correctly. It is possible, but is not consistent with our data, at least for the domain 4 S4 segment, because when you rotate from one side to another there are some residues that are on one side of the electric field, say intracellular, and then you are rotating them to an extracellular vestibule on the other side. This means that the residues on the back side are moving in the opposite direction. For example, our residues R2 and R3 are on the intracellular side at hyperpolarized voltages. The two neutral residues in between would have to be on the extracellular side. Then they should have to move in opposite directions. It turns out that those two residues are on the same side. This rules out pure rotation, at least for that S4 segment.

Strichartz: You speak about the electric field like it is an invariant, but it is the gradient of the voltage. You are getting a dielectric relaxation across the gating region when the thing turns, so that the distribution of the driving force is changing during the very process of gating.

Horn: There isn't a pure hydrophobic slab with charges moving across it. There are some data from Fred Sigworth's lab showing that these vestibules are not connected to the intracellular surface in terms of electric field (Islas & Sigworth 2001). The electric field partly moves through those crevices the way a charged blocker will move through the electric field to get to a binding site. They are narrow enough that you don't start out hydrophilic, then hit the electric field, and then arrive on the other side: it is a gradual movement. You are right: during movement, things can change.

Strichartz: They must change.

References

Featherstone DE, Fujimoto E, Ruben PC 1998 A defect in skeletal muscle sodium channel deactivation exacerbates hyperexcitability in human paramyotonia congenita. J Physiol 506:627–638

Islas LD, Sigworth FJ 2001 Electrostatics and the gating pore of *Shaker* potassium channels. J Gen Physiol 117:69–89

Isom LL 2002 β subunits: players in neuronal hyperexcitability? In: Sodium channels and neuronal hyperexcitability. Wiley, Chichester (Novartis Found Symp 241) p 124–143

Diverse functions and dynamic expression of neuronal sodium channels

Stephen G. Waxman, Theodore R. Cummins, Joel A. Black and Sulayman Dib-Hajj

Department of Neurology and PVA/EPVA Neuroscience Research Center, Yale School of Medicine, New Haven, CT 06510, and Rehabilitation Research Center, VA Connecticut, West Haven CT 06516, USA

Abstract. Nearly a dozen genes encode different Na^+ channels, sharing a common overall motif but with subtly different amino acid sequences. Physiological signatures have now been established for some Na^+ channels and it is clear that, from a functional point of view, Na^+ channels are not all the same: different channels can have different physiological characteristics, and they can play different roles in the physiology of excitable cells. Moreover, the expression of Na^+ channels within neurons is not a static process. Plasticity of Na^+ channel gene expression occurs in the normal nervous system, where it accompanies transitions between different physiological states (e.g. low-frequency versus high-frequency firing states) in some types of neurons. Maladaptive changes in Na^+ channel gene expression also occur in some pathological neurons. For example, transection of the peripheral axons of spinal sensory neurons triggers down-regulation of some Na^+ channel genes and up-regulation of others, resulting in changes in Na^+ current expression that produce hyperexcitability, thereby contributing to chronic pain. There is also recent evidence for the expression of normally silent Na^+ channel genes in Purkinje cells in experimental models of demyelinating diseases and in a human disease, multiple sclerosis; this dysregulation of Na^+ channel expression may interfere with neuronal function in these disorders. The diversity and dynamic nature of Na^+ channel expression introduce a high degree of complexity into the nervous system and present challenges for neuroscientists. In addition, they may present therapeutic opportunities as selective modulators for various Na^+ channel subtypes become available.

2002 Sodium channels and neuronal hyperexcitability. Wiley, Chichester (Novartis Foundation Symposium 241) p 34–60

Elucidation of the function, and underlying molecular structure, of Na^+ channels is one of the triumphs of modern neurobiology. The pivotal studies of Hodgkin & Huxley (1952) taught us that, in most neurons, voltage-gated Na^+ channels produce the regenerative depolarization that underlies the action potential. Although these pioneering neuroscientists were not able to visualize Na^+ channels or directly discern their molecular structure, they were able to infer some aspects of the structure of these channels, and to predict, for example, that they possess 'gates' which opened as the channels were stimulated, allowing Na^+ ions to pass through. Decades later, the molecular revolution permitted the cloning of the first Na^+ channel by Numa and his colleagues (Noda et al 1984).

There were hints, from prior electrophysiological and pharmacological experiments, that there might be more than one type of voltage-gated neuronal Na^+ channel. However, the idea that there are multiple neuronal Na^+ channels had its formal birth when Numa and his colleagues demonstrated the expression, within the mammalian brain, of three Na^+ channel genes, each encoding a different but related molecule (Noda et al 1986). Following this, other genes encoding additional Na^+ channels were discovered. We now know that there are, in fact, nearly a dozen genes, each encoding a Na^+ channel with a subtly different amino acid sequence. Physiological signatures have now been established for some of these Na^+ channels, and it is clear that, from a functional point of view, Na^+ channels are not all the same: different channels can have different physiological characteristics, and they can play different roles in the physiology of excitable cells.

It has also become clear that the expression of Na^+ channels within neurons is not a static process. Although neuronal plasticity has been most extensively studied in terms of synaptic potentiation and depression, sprouting and pruning of neurites, and the recruitment of pre-existing or new neurons into functional circuits, there is now abundant evidence for plasticity of the neuronal electrogenic apparatus: the expression of genes encoding voltage-gated Na^+ channels is dynamic.

This chapter will first review recent progress on the diverse physiological properties and functions of different Na^+ channels. It will then discuss dynamic aspects of Na^+ channel expression, both in normal neurons and in diseased neurons.

More than fast threshold devices: Na^+ channels can function as amplifiers of slow subthreshold depolarizations

Although the concept of the 'fast' transiently-active Na^+ channel as a mediator of the depolarizing phase of the action potential is now regarded as classical, it has also become apparent that some Na^+ channels can function within a voltage domain that is subthreshold for action potential generation so that they act as boosters of subthreshold depolarizing inputs. An example is provided by the PN1 Na^+ channel, which is selectively and prominently expressed within dorsal root ganglion (DRG) neurons (Toledo-Aral et al 1997). To establish the physiological role of PN1, Cummins et al (1998) used patch-clamp techniques to study its human ontology PN1/hNE expressed in a heterologous expression system (HEK293 cells), in a 'bottom-up' analysis. The physiological signature of PN1/hNE channels is displayed in Fig. 1 B–H. For comparison, Fig. 1A shows patch-clamp

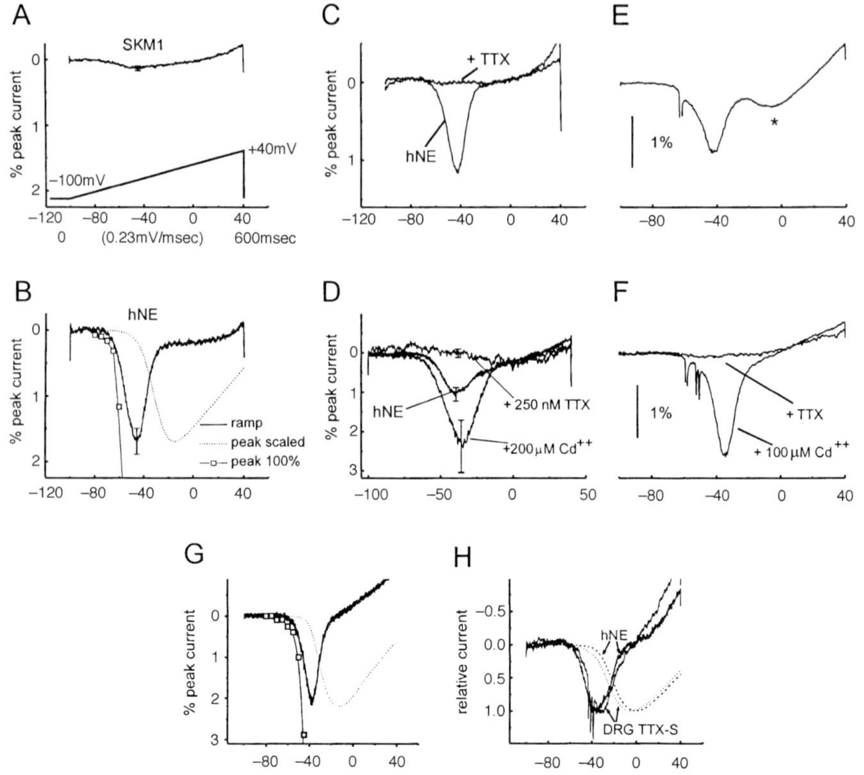

FIG. 1. 'Bottom-up' and 'top-down' analyses reveal similar properties of the PN1/hNE Na^+ channel in HEK293 cells and in DRG neurons. (A) SkM1 Na^+ channels, transfected into HEK293 cells, do not activate in response to slow ramp-like (0.23 mV/ms) depolarizations. (B) In contrast, these slow ramp stimuli activate PN1/hNE channels transfected into HEK293 cells, generating distinct inward currents which are evoked close to resting potential. (C) PN1/hNE currents are blocked by TTX. (D) PN1/hNE currents are enhanced by Cd^{2+}. (E) In a 'top-down' analysis, identical ramp stimuli evoke a similar inward current in DRG neurons. (F) Ramp currents in DRG neurons display a similar pharmacologic profile, being blocked by TTX and enhanced by Cd^{2+}. (G) The threshold for activation of ramp currents within intact DRG neurons is similar to that for isolated PN1/hNE currents. (H) PN1/hNE currents in HEK293 cells, and ramp currents in intact DRG neurons, plotted together to facilitate comparison of their voltage-dependences. Note that they overlie each other. Modified from Cummins et al (1998).

recordings from SkM1 (muscle) Na^+ channels, transfected into HEK293 cells. Like most traditional Na^+ channels, the SkM1 channels require sudden, relatively large depolarizations in order to be activated. In response to slow depolarizations close to resting potential, these channels do not open; when stimulated, for example, by slow ramp-like stimuli (0.23 mV/ms), the SkM1 channels do not generate a response (Fig. 1A). PN1/hNE channels display slow closed-state inactivation and recovery from inactivation. As a result of the slow recovery from inactivation by PN1/hNE channels, cells expressing these channels are not capable of firing at the high frequencies that can be reached by cells expressing SKM1 channels. The presence of PN1/hNE channels does, however, confer a functional advantage. Closed-state inactivation develops slowly in PN1/hNE channels, up to fivefold slower than in SkM1 channels (Cummins et al 1998). Thus, in contrast to SkM1, PN1/hNE channels activate and generate a Na^+ current in response to slow ramp-like stimuli, even when small depolarizations close to resting potential are used as stimuli (Fig. 1B). The activity of the PN1/hNE channels also exhibits a unique physiological profile, being blocked by tetrodotoxin (Fig. 1C) and enhanced by cadmium (Fig. 1D).

Having established the functional signature for PN1/hNE channels with this bottom-up approach, Cummins et al (1998) next used this information to design a top-down approach, to determine whether these channels play a similar role in their native environment within DRG neurons. Because ramp-like depolarizations constitute an effective stimulus for the PN1/hNE Na^+ channel in HEK293 cells, depolarizing ramps were used in the stimulus protocol for intact DRG neurons, and were found to evoke depolarizing responses within these cells (Fig. 1E), similar to those seen in the HEK expression system (Fig. 1B). The Na^+ current elicited by these ramp stimuli within intact DRG neurons displayed responses to TTX and cadmium (Fig. 1F) that are identical to those of PN1/hNE channels studied in the HEK293 expression system. As seen in Fig. 1H, moreover, the voltage-dependence of the TTX-sensitive ramp current in DRG neurons is nearly identical to that of the PN1/hNE current in HEK cells. The bottom-up and top-down analyses thus converge and demonstrate that, in intact DRG neurons, the PN1/hNE channel responds to small, slow depolarizations close to resting potential, activating so as to produce inward (depolarizing) currents. These functional properties suggest that PN1/hNE channels respond to small depolarizing inputs, acting as boosters that amplify them. PN1/hNE channels are localized at the distal ends of neurites of spinal sensory neurons in culture (Toledo-Aral et al 1997); if these channels were similarly deployed *in vivo*, their localization would place them close to the sites of generator potential activity in nerve terminals.

Na^+ channels can contribute to resting potential and modulate neuronal excitability

Dib-Hajj et al (1998a) cloned and sequenced another channel, NaN (also termed SNS 2 by Tate et al 1998), which is preferentially expressed in small DRG neurons. The presence of a serine at position 355 in NaN suggests that it is TTX-resistant (Dib-Hajj et al 1998a). The gene for NaN is localized, together with genes

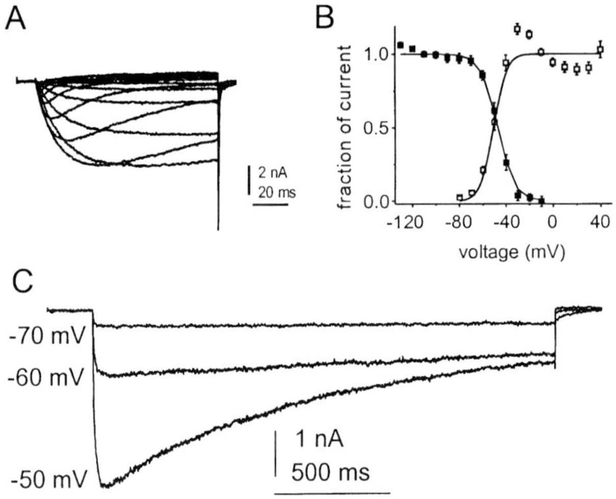

FIG. 2. Persistent TTX-resistant Na^+ currents are produced by NaN channels in small DRG neurons. (A) Representative TTX- resistant Na^+ currents recorded from a DRG neuron from a SNS-null mouse with 100 ms test pulses. (B) Activation (unfilled squares) and steady-state inactivation (filled squares) curves for the NaN current show significant overlap. Steady-state inactivation was measured in SNS-null neurons with 500 ms prepulses. (C) NaN currents from an SNS-null neuron, elicited with 2 s step depolarizations to the voltages indicated. Recordings were made with 250 nM TTX, 100 μM cadmium (to block Ca^{2+} currents) and $V_{hold} = -120$ mV in an SNS-null neuron. Modified from Cummins et al (1999).

encoding two other TTX-resistant Na^+ channels, SNS and hSkM2, within a conserved linkage group at 3p21-3p24 within human chromosome 3 (Dib-Hajj et al 1999a). NaN and SNS are the only TTX-resistant Na^+ channels that are present within DRG neurons. Thus, SNS-knockout mutant mice (Akopian et al 1999) provide a model system in which NaN can be studied in isolation. Cummins et al (1999) observed a TTX-resistant ($K_i = 39$ μM) persistent Na^+ current (Fig. 2A, C) attributable to NaN in SNS-knockout DRG neurons. A similar TTX-resistant persistent current, which appears to be produced by NaN, can be recorded from wild-type mouse and rat DRG neurons (Cummins et al 1999) and human DRG neurons (Dib-Hajj et al 1999b). The persistent current is unique in exhibiting a hyperpolarized dependence of activation (threshold −70 mV; midpoint of activation = −41 mV in mouse DRG neurons) and steady-state inactivation (midpoint = −44 mV), with a substantial overlap between activation and steady-state inactivation curves (Fig. 2B) which extends from −70 mV to −30 mV (Cummins et al 1999). Since the resting potential of small DRG neurons is close to −55 mV (Caffrey et al 1992), these results suggest that some NaN channels should be active, generating a 'window current', near resting membrane potential.

Moreover, the low threshold for activation of NaN channels suggests that they should open in response to small subthreshold depolarizations close to membrane potential, thus contributing to subthreshold electrogenesis and regulating the excitability of DRG neurons. Computer simulations suggest that NaN channels contribute a 10 mV to 15 mV depolarizing influence to resting potential, and amplify small depolarizing inputs by more than 50%, in small DRG neurons (Herzog et al 2001).

Na^+ channel gene expression changes as neurons pass from one functional state to another

It is of course well known that, as neurons pass from a quiescent state to a high-frequency firing state, they use their (pre-existing) Na^+ channels differently, i.e. they activate these channels repetitively. But do neurons produce a different set of channels when they make these state transitions? To determine whether Na^+ channel gene expression changes as neurons pass from a low- to a high-frequency firing state, Tanaka et al (1999) used magnocellular neurosecretory cells within the supraoptic nucleus as a model. These specialized neurons send their axons to the neurohypophysis where they release vasopressin after becoming active in response to increases in plasma osmolality. In their basal state these cells are relatively quiescent, firing irregularly at low frequencies (<3 impulses/s) but, in response to changes in the osmotic milieu, these cells respond by generating high-frequency bursts of action potentials which trigger the release of vasopressin (Li & Hatton 1996). The hypothesis, that the transition from the quiescent to the bursting state is accompanied by a change in Na^+ channel gene expression, was tested by studying magnocellular neurons under normal conditions and following salt-loading, which triggers a transition to a bursting state. *In situ* hybridization was used to measure the transcription of mRNA for the various Na^+ channels. Notably, these studies revealed that salt loading is accompanied by a distinct up-regulation of mRNAs for two Na^+ channels, α-II and Na6, in these neurons (Fig. 3).

Because the expression of ion channels within the cell membrane is controlled at both the transcriptional and translational levels, increased mRNA levels are not necessarily accompanied by increased synthesis of channel protein. As a next step, Tanaka et al (1999) thus studied control and salt-loaded magnocellular neurons using immunocytochemical and immunoblotting methods with an antibody directed against a conserved region of Na^+ channels. An increase in the level of Na^+ channel protein within the magnocellular neurons of salt-loaded rats, showed that the changes in Na^+ channel gene transcription were paralleled by increases in Na^+ channel protein.

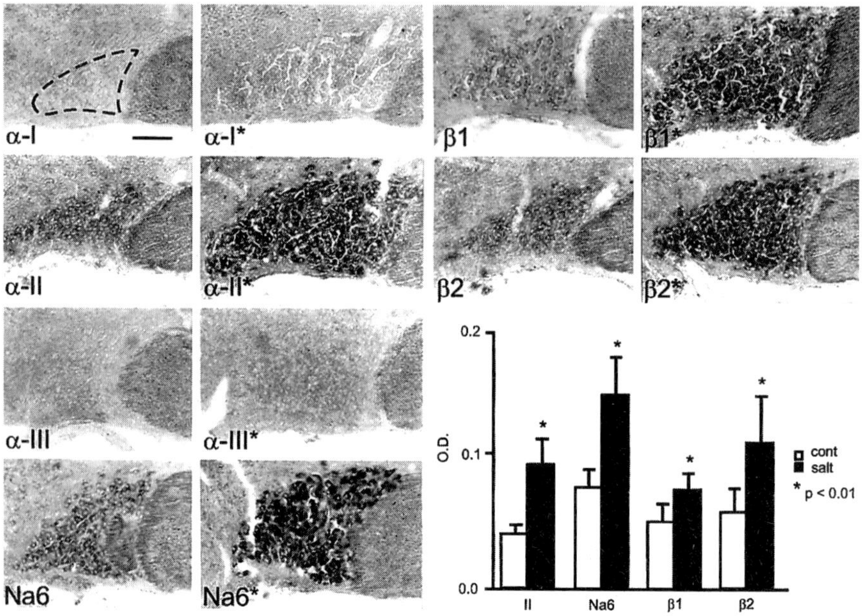

FIG. 3. α-II and Na6 Na^+ channel mRNAs are up-regulated, together with Na^+ channel $\beta1$ and $\beta2$ mRNA, in supraoptic magnocellular neurons following salt loading. The micrographs, from control (left column) and salt-loaded (right column) rats, were digitally enhanced to show *in situ* hybridization with subtype-specific riboprobes for Na^+ channel subunits α-I, α-II, α-III and Na6. α-I and α-III mRNA are not detectable, and low levels of α-II and Na6 mRNA are present in the control supraoptic nucleus (no asterisks). Expression of the α-II and Na6 transcripts is up-regulated following salt-loading (asterisks). Optical densities from unenhanced micrographs (graph, lower right) provide a quantitative measure of mRNA levels and show a significant increase in α-II and Na6 mRNA following salt loading. $*P<0.01$. Bar $=100\,\mu$m. Modified from Tanaka et al (1999).

Having demonstrated changes in Na^+ channel mRNA and protein expression as magnocellular neurons made the transition from one functional state to another, the next question was whether these *molecular* changes were associated with *functional* changes in these neurons. To answer this question, Tanaka et al (1999) used patch-clamp recording to study these cells. The presence of two different Na^+ channels (α-II and Na6) in the magnocellular neurons suggested that these cells should produce several distinct Na^+ currents. It was known, from studies on other neuronal cell types such as Purkinje cells, that the Na6 Na^+ channel can produce a slow or persistent Na^+ current (Vega-Saenz de Miera et al 1997, Raman et al 1997). The α-II channel, in contrast, has been shown to produce a fast transient current (Noda et al 1986, Auld et al 1988). Consistent with the expression of these two types of Na^+ channels, patch-clamp recordings

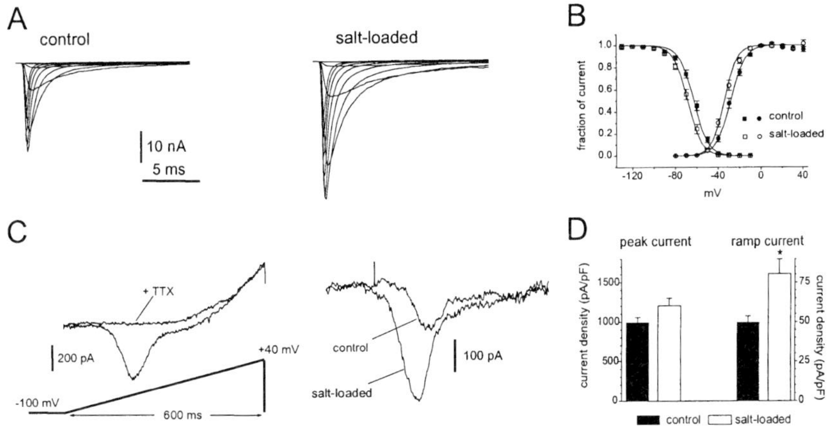

FIG. 4. Two distinct Na$^+$ currents in supraoptic neurons show differential increases following salt loading. (A) Recordings from representative supraoptic neurons acutely isolated from control (left panel) or salt-loaded (right panel) rats, showing the fast transient Na$^+$ current. The currents were elicited by 40 ms test pulses to various potentials from −60 to 30 mV. Cells were held at −100 mV. (B) Normalized activation (circles) and steady-state inactivation (squares) curves show only small differences between control (filled symbols) and salt-loaded (open symbols) neurons. Error bars indicate SEM. (C) Ramp currents are elicited by slow ramp-like depolarizations (extending from −100 to +40 mV over 600 ms) in supraoptic neurons. The left panel shows that TTX (250 nM) blocks the ramp current in salt-loaded supraoptic neurons, demonstrating that this current is produced by Na$^+$ channels. The right panel shows the TTX-sensitive ramp currents in representative control and salt-loaded supraoptic neurons. Note the larger amplitude in salt-loaded neurons. (D) Fast transient and ramp current densities (estimated by dividing the maximum currents by the cell capacitance) are both increased following salt-loading; however, the increase is proportionately greater for the ramp currents. Error bars indicate SEM, *indicates $P < 0.005$. From Tanaka et al (1999).

demonstrated two distinct Na$^+$ currents in control magnocellular neurons: first, a fast transient Na$^+$ current which would be expected to contribute to the rapid depolarizing upstroke of the action potential (Fig. 4A); and second, a persistent 'threshold' Na$^+$ current which activates closer to resting potential (Fig. 4C). The patch-clamp experiments showed that there was an increase of 20% in the density of the fast transient Na$^+$ current in salt-loaded rats. In contrast, the density of the persistent threshold current was increased by approximately 60% (Fig. 4D). The two Na$^+$ currents were increased to significantly different degrees. The disproportional increases in the two currents encoded by these two channels would be expected to lower the threshold for action potential generation.

Magnocellular neurons represent an example of molecular and functional remodelling of the neuronal electrogenic apparatus in the normal nervous system. As these cells move from a quiescent to a bursting state, they alter their pattern of Na^+ channel gene activation, a change that alters the threshold of these neurons.

Expression of some Na^+ channel genes is down-regulated in injured neurons

The striking plasticity in Na^+ channel expression that can occur in normal neurons raises the question of whether there are alterations in Na^+ channel expression in injured neurons. Dorsal root ganglion (DRG) neurons have been especially well studied in this respect. As illustrated in Fig. 5 (middle and bottom rows), following transection of the peripheral axons of DRG neurons (by ligation of the sciatic nerve), expression of mRNA for the SNS (Dib-Hajj et al 1996) and NaN (Dib-Hajj et al 1998a) Na^+ channels is down-regulated. The reductions in SNS and NaN transcript expression are paralleled by reduced levels of SNS and NaN Na^+ channel protein (Sleeper et al 2000).

While transection of the peripheral axons of DRG neurons triggers a down-regulation of SNS and NaN, transection of the centrally-direct branches of DRG neurons (dorsal rhizotomy) does not alter the expression of these channels (Sleeper et al 2000). These results suggest that down-regulation of SNS and NaN might be due to loss of access to trophic factors produced by peripheral targets. Consistent with this hypothesis, nerve growth factor (NGF) up-regulates SNS expression (Black et al 1997), while glial derived neurotrophic factor (GDNF) up-regulates NaN and SNS expression (Fjell et al 1999) in DRG neurons *in vitro*. Similarly, delivery of NGF (Dib-Hajj et al 1998b) and GDNF (Cummins et al 2000) to the injured nerve stump *in vivo* can rescue the expression of SNS and NaN, respectively.

In parallel with the down-regulation of SNS and NaN mRNA and protein, the slowly-inactivating and persistent TTX-resistant Na^+ currents produced by these channels are reduced in axotomized neurons (Cummins & Waxman 1997, Sleeper et al 2000) (Fig. 6A, B). These physiological changes persist for at least 60 days post-axotomy (Cummins & Waxman 1997). Loss of NaN (persistent TTX-resistant) channels would be expected to shift resting potential in a hyperpolarizing direction, reducing resting inactivation of fast Na^+ channels, and thereby producing DRG neuron hyperexcitability which can contribute to pain and parasthesia (Cummins & Waxman 1997). Consistent with a contribution of these changes to the pathogenesis of neuropathic pain, Dib-Hajj et al (1999c) observed a similar (although quantitatively smaller) down-regulation in SNS and NaN channels and their currents in DRG neurons in the chronic constriction injury model of neuropathic pain.

The previously silent type III Na^+ channel gene is expressed in axotomized dorsal root ganglion neurons

Early electrophysiological studies (Eccles et al 1958) demonstrated that, following axonal transection, there are changes in somatodendritic excitability which appear to reflect the deployment of increased numbers of Na^+ channels within the neuronal membrane. More recent electrophysiological studies have confirmed these findings (Kuno & Llinas 1970) and have established that this abnormal somatodendritic excitability is Na^+-dependent (Sernagor et al 1986, Titmus & Faber 1986). Immunocytochemical studies with antibodies that do not distinguish Na^+ channel subtypes have demonstrated abnormal accumulations of Na^+ channels at injured axonal tips within experimental neuromas (Devor et al 1989, England et al 1994, 1996) and it has been suggested that these changes are a result of a change in vectorial transport of Na^+ channels (Titmus & Faber 1990, Devor 1994). Although these studies demonstrated that there can be an increased *number* of Na^+ channels within injured neurons, they did not address the question of whether abnormal *types* of Na^+ channels are produced in injured neurons.

Waxman et al (1994) and Dib-Hajj et al (1996) used *in situ* hybridization and RT-PCR to show that there is a striking up-regulation of the previously silent α-III Na^+ channel gene in DRG neurons following transection of their axons (Fig. 5, top). Iwahashi et al (1994) observed a similar up-regulation of the previously silent α-III Na^+ channel gene following axotomy of adult facial motor neurons. These changes are not due to a global increase in channel protein synthesis since, as noted above, other Na^+ channel genes are down-regulated in these axotomized neurons. Black et al (1999a) demonstrated that increased levels of type III Na^+ channel protein accompany the up-regulation of type III mRNA in axotomized DRG neurons.

In parallel with the activation of the previously silent type III Na^+ channel gene in axotomized DRG neurons, there is a discrete switch in the properties of the Na^+ currents that are expressed by these cells (Fig. 6C, D). Specifically, a TTX-sensitive Na^+ current that recovers (reprimes) rapidly from inactivation emerges in axotomized neurons, i.e. there is a switch from a slowly repriming TTX-sensitive Na^+ current to a rapidly-repriming current (Cummins & Waxman 1997). Recovery from inactivation is accelerated by as much as fourfold in axotomized neurons, and this appears to contribute to the hyperexcitability of these injured cells. Although the physiological signature of the type III Na^+ channel has not yet been fully delineated in mammalian systems, it has been suggested that type III channels produce the rapidly-repriming TTX-sensitive current in axotomized neurons (Cummins & Waxman 1997). This proposal is supported by several observations: first, rapidly-repriming TTX-sensitive Na^+ current and type III Na^+ channels show parallel patterns of up-regulation following the transection

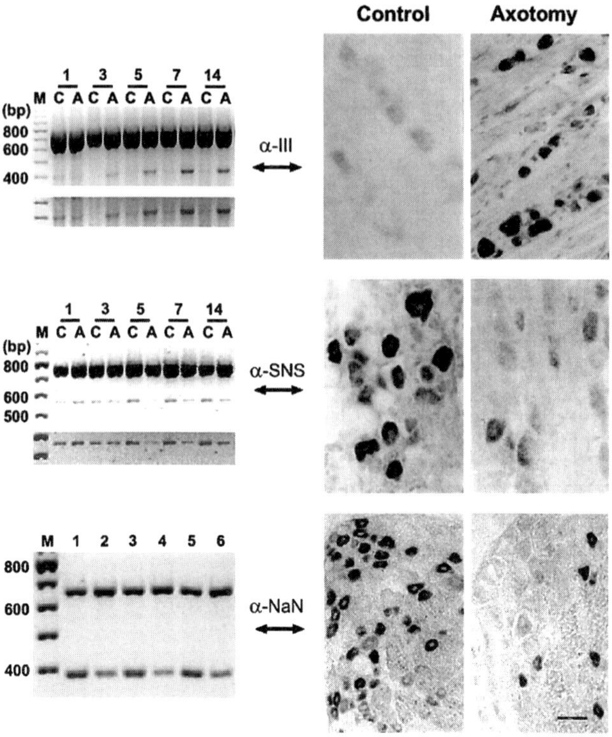

FIG. 5. Na^+ channel expression can change strikingly in neurons following injury. mRNA for Na^+ channel α-III (top) is up-regulated, and mRNA for SNS (middle) and NaN (bottom) are down-regulated, in DRG neurons following transection of their axons within the sciatic nerve. The *in situ* hybridizations (right side) show α-III, SNS and NaN mRNA in control DRG, and at 5–7 days post-axotomy. RT-PCR (left side) shows products of co-amplification of α-III and SNS together with β-actin transcripts in control (C) and axotomized (A) DRG (days post-axotomy indicated above gels), with computer-enhanced images of amplification products shown below gels. Co-amplification of NaN (392 bp) and GAPDH (6076 bp) shows decreased expression of NaN mRNA at 7 days post-axotomy (lanes 2,4,6) compared to Controls (lanes 1, 3, 5). Top and middle panels modified from Dib-Hajj et al (1996). Bottom modified from Dib-Hajj et al (1998a).

of peripherally directed (sciatic nerve) axons within DRG neurons but not following transection of the centrally-directed (dorsal root) axons of these cells (Black et al 1999a). Second, type III Na^+ channels display rapid repriming when expressed in mammalian expression systems (HEK293 cells; T.R. Cummins, S.D. Dib-Hajj & S.G. Waxman, unpublished work). Third, abnormal accumulations of type III Na^+ channel proteins can be detected close to the tips of injured axons within experimental neuromas (Black et al 1999a), a site where abnormal hyperexcitability has been demonstrated (Scadding 1981, Burchiel 1984, Matzner

& Devor 1992). It appears likely that turning on of the previously silent type III Na^+ channel gene contributes to the development of hyperexcitability of axotomized DRG neurons.

The previously silent SNS Na^+ channel gene is abnormally expressed in some demyelinated neurons

The *S*ensory *N*euron *S*pecific Na^+ channel SNS (Akopian et al 1996, Sangameswaran et al 1996) is normally expressed in a highly specific manner within primary sensory neurons within DRG and trigeminal ganglion, and is not present within healthy neurons within the uninjured brain. This sensory neuron-specific TTX-resistant Na^+ channel exhibits slow activation and inactivation kinetics and a depolarized voltage-dependence (Akopian et al 1996, Sangameswaran et al 1996), and more rapid recovery from inactivation (Elliott & Elliott 1993, Dib-Hajj et al 1997) than traditional 'fast' Na^+ channels. As a result of the distinct electrophysiological characteristics of SNS-type channels, the presence of these channels can alter the firing properties of neurons (Akopian et al 1999, Schild & Kunze 1997).

Multiple sclerosis (MS) has traditionally been viewed as a disorder in which myelin is the primary target. However, there is recent evidence for abnormal SNS expression in experimental models of demyelination and in MS. Black et al (1999b) studied Na^+ channel expression in the taiep rat, a mutant model in which myelin is initially formed normally, but then lost as a result of an oligodendrocyte abnormality. They observed the abnormal expression of SNS Na^+ channel mRNA and protein in Purkinje cells following loss of myelin. More recently, Black et al (2000) demonstrated that SNS mRNA and protein, which are not detectable in normal Purkinje cells, are expressed within Purkinje cells in a mouse model of MS, chronic relapsing experimental allergic encephalomyelitis. Black et al (2000) have also demonstrated the expression of SNS mRNA (Fig. 7a, b) and protein (Fig. 7e, f) within cerebellar Purkinje cells from tissue obtained post-mortem from MS patients, but not in controls with no neurological disease (Fig. 7c, g).

The presence of abnormally expressed SNS Na^+ channels within Purkinje cells in experimental and human 'demyelinating' disorders may have significant functional consequences. Non-pathological Purkinje cells produce multiple Na^+ currents which interact to determine the firing properties of these cells (Llinas & Sugimori 1980, Raman & Bean 1997). Mutations of the Na^+ channels that are normally expressed in Purkinje cells produce altered firing behaviour, and this appears to provide a basis for cerebellar ataxia in mutant models such as jolting mice (Kohrman et al 1996, Raman et al 1997). Normal cerebellar functioning depends on precise timing of impulses. We have suggested that aberrant expression of SNS Na^+ channels within Purkinje cells in the demyelinating

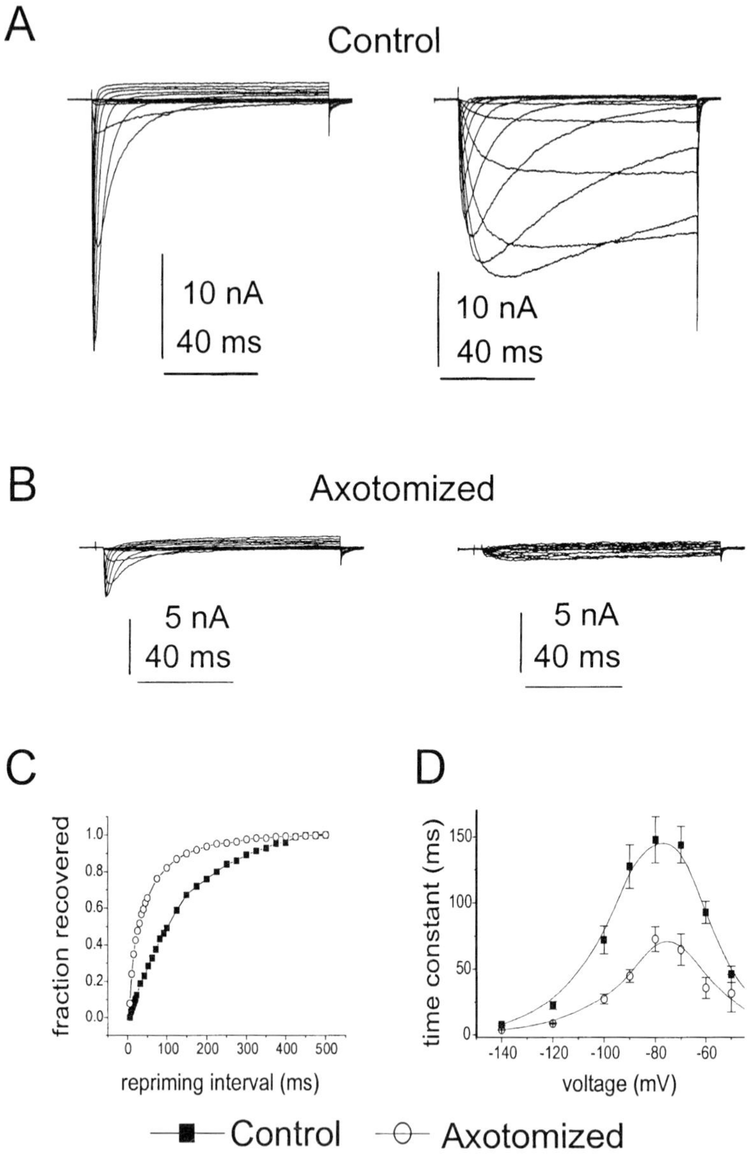
A
Control
10 nA
40 ms
10 nA
40 ms
B
Axotomized
5 nA
40 ms
5 nA
40 ms
C
fraction recovered
1.0
0.8
0.6
0.4
0.2
0.0
0
100
200
300
400
500
repriming interval (ms)
D
time constant (ms)
150
100
50
0
-140
-120
-100
-80
-60
voltage (mV)
Control
Axotomized

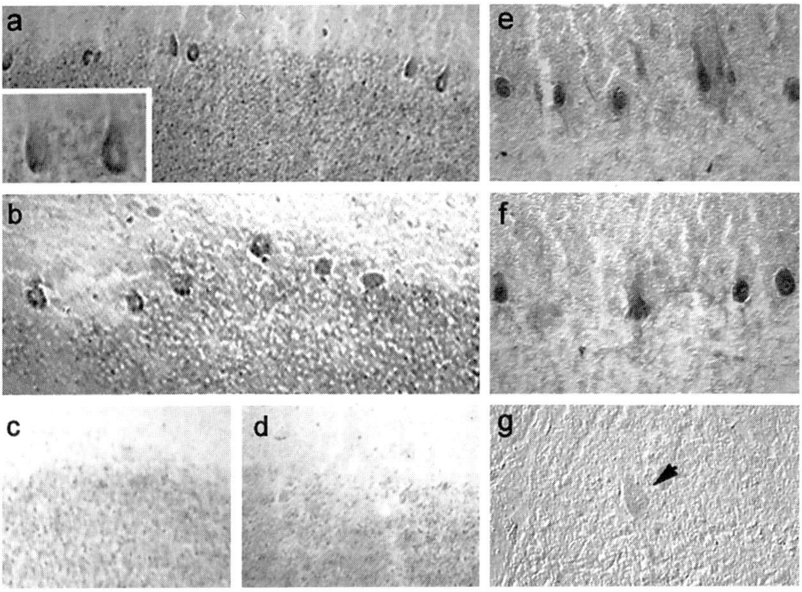

FIG. 7. SNS Na^+ channel is not present within the normal brain, but is expressed in cerebellar Purkinje cells within brains obtained at post-mortem from MS patients. Panels on left show *in situ* hybridization with SNS-specific antisense riboprobes, and demonstrate the absence of SNS mRNA in control cerebellum (c) and its presence in Purkinje cells in post-mortem tissue from two patients with MS (a, b). No signal is present following hybridization with sense riboprobe (d). Panels on right show immunostaining with an antibody directed against SNS, and illustrate absence of SNS protein in control cerebellum (g, arrow indicates Purkinje cell) and its presence in MS (e, f). Modified from Black et al (2000).

FIG. 6. (*opposite*) In parallel with down-regulation of SNS and NaN channels, slow and persistent TTX-resistant Na^+ currents in small DRG neurons are down-regulated following axonal transection within the sciatic nerve. (A) patch clamp recordings showing SNS (left) and NaN (right) currents from representative control DRG neurons. (B) SNS and NaN currents are reduced in axotomized (B, 6 days post-axotomy) DRG neurons. (Modified from Cummins & Waxman 1997.) (C) A rapidly repriming TTX-sensitive Na^+ current emerges in DRG neurons following axonal injury. The graph displays the time course for recovery from inactivation at -80 mV for the peak currents, for control (filled circles) and axotomized (open circles, 7 days post-axotomy). Recovery is much faster for the currents in the axotomized neuron. Single exponential functions fitted to the data gave time constants of 160 ms for the control neuron and 41 ms for the axotomized neuron. (D) Time constants for recovery from inactivation for control neurons (closed squares) and peripherally axotomized neurons (open circles), plotted as a function of voltage. Cells were prepulsed to -20 mV for 20 ms to inactivate all of the current, then brought back to the indicated recovery voltage for increasing recovery durations prior to a test pulse to -20 mV. Time constants were estimated from single exponential fits. Repriming is significantly accelerated at all voltages between -60 and $=100$ mV. (Modified from Cummins & Waxman 1997 and Black et al 1999.)

disorders may alter the firing pattern of these neurons, a form of molecular mistuning that might be expected to contribute to clinical abnormalities such as ataxia (Black et al 2000).

Diversity and dynamic expression: challenges and opportunities

The multiplicity and diversity of Na^+ channels, and the fact that their expression is dynamic rather than fixed, underscore the complexity of neurons and the electrogenic machinery that they house. Differences in the ensemble of Na^+ channels produced by different types of neurons, and state-dependent changes in Na^+ channel expression present daunting challenges to investigators. However, now that the open reading frames for most of these channels have been elucidated, it is possible to study their expression at the mRNA level with great precision. Moreover, an increasing number of subtype-specific antibodies are becoming available. New subtype-specific toxins are being discovered and additional subtype-specific blockers may soon be developed. These will hasten the elucidation of the physiological properties and functional roles of the various Na^+ channel subtypes, so that it may soon be possible to dissect the electrogenic machinery within any given type of neuron with unprecedented precision. The complexity of Na^+ channel gene expression in neurons may also present some therapeutic opportunities. In so far as abnormal Na^+ channel expression contributes to hypo- and hyperexcitability and/or distorted impulse trafficking, the development of subtype-specific Na^+ channel modulators may provide new therapeutic inroads for the treatment of diseases of the nervous system in which neuronal excitability is altered.

Acknowledgements

Research described in this chapter has been supported, in part, by grants from the National Multiple Sclerosis Society, and from the Rehabilitation Research Service and Medical Research Service, Department of Veterans Affairs. We also thank the Eastern Paralysed Veterans Association and the Paralysed Veterans of America for support, including a gift that supports the Yale–London Collaboration.

References

Akopian AN, Sivilotti L, Wood JN 1996 A tetrodotoxin-resistant voltage-gated sodium channel expressed by sensory neurons. Nature 379:257–262

Akopian AN, Souslova V, England S et al 1999 The tetrodotoxin-resistant sodium channel SNS has a specialized function in pain pathways. Nat Neurosci 2:541–548

Auld VJ, Goldin AL, Krafte DS et al 1988 A rat brain Na^+ channel alpha subunit with novel gating properties. Neuron 1:449–461

Black JA, Langworthy K, Hinson AW, Dibb-Hajj SD, Waxman SG 1997 NGF has opposing effects on Na^+ channel III and SNS gene expression in spinal sensory neurons. Neuroreport 8:2331–2335

Black JA, Cummins TR, Plumpton C et al 1999a Upregulation of a silent sodium channel after peripheral, but not central, nerve injury in DRG neurons. J Neurophysiol 82:2776–2785

Black JA, Fjell J, Dibb-Hajj S et al 1999b Abnormal expression of SNS/PN3 sodium channel in cerebellar Purkinje cells following loss of myelin in the *taiep* rat. Neuroreport 10:913–918

Black JA, Dib-Hajj S, Baker D, Newcombe J, Cuzner ML, Waxman SG 2000 Sensory Neuron Specific sodium channel SNS is abnormally expressed in the brains of mice with experimental allergic encephalomyelitis and humans with multiple sclerosis. Proc Natl Acad Sci USA 97:11598–11602

Burchiel KJ 1984 Effects of electrical and mechanical stimulation on two foci of spontaneous activity which develop in primary afferent neurons after peripheral axotomy. Pain 18:249–265

Caffrey JM, Eng DL, Black JA, Waxman SG, Kocsis JD 1992 Three types of sodium channels in adult rat dorsal root ganglion neurons. Brain Res 592:283–297

Cummins TR, Waxman SG 1997 Downregulation of tetrodotoxin-resistant sodium currents and upregulation of a rapidly repriming tetrodotoxin-sensitive sodium current in small spinal sensory neurons after nerve injury. J Neurosci 17:3503–3514

Cummins TR, Howe JR, Waxman SG 1998 Slow closed-state inactivation: a novel mechanism underlying ramp currents in cells expressing the hNE/PN1 sodium channel. J Neurosci 18:9607–9619

Cummins TR, Dib-Hajj SD, Black JA, Akopian AN, Wood JN, Waxman SG 1999 A novel persistent tetrodotoxin-resistant sodium current in SNS-null and wild-type small primary sensory neurons. J Neurosci 19:1–6

Cummins TR, Black JA, Dib-Hajj SD, Waxman SG 2000 GDNF up-regulates expression of functional SNS and NaN sodium channels and their currents in axotomized DRG neurons. J Neurosci 20:8754–8762

Devor M 1994 The pathophysiology of damaged peripheral nerves. In: Wall PD, Melzack R (eds) Textbook of Pain. Churchill Livingstone, New York, p 79–89

Devor M, Keller CH, Deerinck TJ, Levinson SR, Ellisman MH 1989 Na^+ channel accumulation on axolemma of afferent endings in nerve end neuromas in *Apteronotus*. Neurosc Lett 102:149–154

Dib-Hajj SD, Black JA, Felts P, Waxman SG 1996 Sodium channel α subunit mRNAs I, II, III, NaG, Na6 and NHE: different expression patterns in developing rat nervous system. Molec Brain Res 45:71–83

Dib-Hajj SD, Ishikawa K, Cummins TR, Waxman SG 1997 Insertion of a SNS-specific tetrapeptide in the S3-S4 linker of D4 accelerates recovery from inactivation of skeletal muscle voltage-gated Na channel μ1 in HEK293 cells. FEBS Lett 416:11–14

Dib-Hajj SD, Tyrrell L, Black JA, Waxman SG 1998a NaN, a novel voltage-gated Na channel preferentially expressed in peripheral sensory neurons and down-regulated after axotomy. Proc Natl Acad Sci USA 95:8963–8968

Dib-Hajj SD, Black JA, Cummins TR, Kenney AM, Kocsis JD, Waxman SG 1998b Rescue of alpha-SNS sodium channel expression in small dorsal root ganglion neurons after axotomy by nerve growth factor *in vivo*. J Neurophysiol 79:2668–2676

Dib-Hajj SD, Tyrrell L, Escayg A, Wood PM, Meisler MH, Waxman SG 1999a Coding sequence, genomic organization, and conserved chromosomal localization of mouse gene *Scn11a* encoding the sodium channel NaN. Genomics 59:309–318

Dib-Hajj SD, Tyrrell L, Cummins TR, Black JA, Wood PM, Waxman SG 1999b Two tetrodotoxin-resistant sodium channels in human dorsal root ganglion neurons. FEBS Lett 462:117–120

Dib-Hajj SD, Fjell J, Cummins TR 1999c Plasticity of sodium channel expression in DRG neurons in the chronic constriction injury model of neuropathic pain. Pain 83:591–600

Eccles JC, Libet B, Young RR 1958 The behavior of chromatolysed motoneurons studied by intracellular recording. J Physiol 143:11–40

Elliott AA, Elliott JR 1993 Characterization of TTX-sensitive and TTX-resistant sodium currents in small cells from adult rat dorsal root ganglia. J Physiol 463:39–56

England JD, Gamboni F, Ferguson MA, Levinson SR 1994 Sodium channels accumulate at the tips of injured axons. Muscle Nerve 17:593–598

England JD, Happel LT, Kline DG et al 1996 Sodium channel accumulation in humans with painful neuromas. Neurology 47:272–276

Fjell J, Cummins TR, Dib-Hajj SD, Fried K, Black JA, Waxman SG 1999 Differential role of GDNF and NGF in the maintenance of two TTX-resistant sodium channels in adult DRG neurons. Brain Res Mol Brain Res 67:267–282

Herzog RI, Cummins TR, Waxman SG 2001 Persistent TTX-resistant Na^+ current affects resting potential and response to depolarization in simulated spinal sensory neurons. J Neurophysiol 86:1351–1364

Hodgkin AL, Huxley AF 1952 A quantitative description of membrane current and its application to conduction and excitation in nerve. J Physiol 117:500–544

Iwahashi Y, Furutama T, Inagaki S, Morita Y, Takagi H 1994 Distinct regulation of sodium channel types I, II and III following nerve transection. Mol Brain Res 22:341–345

Kohrman DC, Smith MR, Goldin AL, Harris J, Meisler MH 1996 A missense mutation in the sodium channel scn8a is responsible for cerebellar ataxia in the mouse mutant jolting. J Neurosci 16:5993–5999

Kuno M, Llinas R 1970 Enhancement of synaptic transmission by dendritic potentials in chromatolysed motoneurons of the cat. J Physiol 210:807–821

Li Z, Hatton GI 1996 Oscillatory bursting of phasically firing rat supraoptic neurones in low-Ca^{2+} medium: Na^+ influx, cytosolic Ca^{2+} and gap junctions. J Physiol 496:379–394

Llinas, R, Sugimori M 1980 Electrophysiological properties of *in vitro* Purkinje cell somata in mammalian cerebellar slices. J Physiol 305:171–195

Matzner O, Devor M 1992 Na^+ conductance and the threshold for repetitive neuronal firing. Brain Res 597:92–98

Noda M, Shimizu S, Tanabe T et al 1984 Primary structure of *Electrophorus electricus* sodium channel deduced from cDNA sequence. Nature 312:121–127

Noda M, Ikeda T, Kayano T et al 1986 Existence of distinct sodium channel messenger RNAs in rat brain. Nature 320:188–192

Raman IM, Bean BP 1997 Resurgent sodium current and action potential formation in dissociated cerebellar Purkinje neurons. J Neurosci 17:4157–4526

Raman IM, Sprunger LK, Meisler MH, Bean BP 1997 Altered subthreshold sodium currents and disrupted firing patterns in Purkinje neurons of Scn8a mutant mice. Neuron 19:881–891

Sangameswaran L, Delgado SG, Fish LM et al 1996 Structure and function of a novel voltage-gated tetrodotoxin-resistant sodium channel specific to sensory neurons. J Biol Chem 271:5953–5956

Scadding JW 1981 Development of ongoing activity, mechanosensitivity, and adrenaline sensitivity in severed peripheral nerve axons. Exp Neurol 73:345–364

Schild JH, Kunze DL 1997 Experimental and modeling study of Na^+ current heterogeneity in rat nodose neurons and its impact on neuronal discharge. J Neurophysiol 78:3198–3209

Sernagor E, Yarom Y, Werman R 1986 Sodium-dependent regenerative responses in dendrites of axotomized motoneurons in the cat. Proc Natl Acad Sci USA 83:7966–7970

Sleeper AA, Cummins TR, Hormuzdiar W et al 2000 Changes in expression of two tetrodotoxin-resistant sodium channels and their currents in dorsal root ganglion neurons following sciatic nerve injury, but not rhizotomy. J Neurosci 20:7279–7289

Tanaka M, Cummins TR, Ishikawa K, Black JA, Ibata Y, Waxman SG 1999 Molecular and functional remodeling of electrogenic membrane of hypothalamic neurons in response to changes in their input. Proc Natl Acad Sci USA 96:1088–1093
Tate S, Benn S, Hick C et al 1998 Two sodium channels contribute to the TTX-R sodium current in primary sensory neurons. Nat Neurosci 1:653–655
Titmus MJ, Faber DS 1986 Altered excitability of goldfish Mauthner cell following axotomy. II. Localization and ionic basis. J Neurophysiol 55:1440–1454
Titmus MJ, Faber DS 1990 Axotomy-induced alterations in the electrophysiological characteristics of neurons. Prog Neurobiol 35:1–51
Toledo-Aral JJ, Moss BL, He ZJ et al 1997 Identification of PN1, a predominant voltage-dependent sodium channel expressed principally in peripheral neurons. Proc Natl Acad Sci USA 94:1527–1532
Vega-Saenz de Miera E, Rudy B, Sugimori M, Llinas R 1997 Molecular characterization of the sodium channel subunits expressed in mammalian cerebellar Purkinje cells. Proc Natl Acad Sci USA 94:7059–7064
Waxman SG, Kocsis JD, Black JA 1994 Type III sodium channel mRNA is expressed in embryonic but not adult spinal sensory neurons, and is reexpressed following axotomy. J Neurophysiol 72:466–471

DISCUSSION

Bevan: You showed very nicely that axotomy reduced the expression of SNS and NaN. I know you have shown both in culture and *in vivo* that if you supply growth factors such as NGF, you can restore the expression. One of the hypotheses of a partial denervation would be that the damaged nerves are no longer able to take up growth factor supplied by peripheral tissues. If that is a rate-limiting step, the supply of growth factors to the intact nerves would be greater. Does the behaviour of intact nerves in these partial nerve damage models show any change? Is expression of channels down-regulated like in the damaged fibres, or are they up-regulated as one might expect with an increased growth factor supply?

Waxman: Are you asking about adjacent DRG neurons whose axons have been spared?

Bevan: If you do a partial cut in the sciatic nerve, or in the Chung model you look at the intact ganglia, what do you see?

Waxman: That is an important question, but I can't answer it. We have put blinders on trying to understand what the axotomy does and have focused on neurons whose axons are transected. We have hints that neighbouring neurons may show changes in gene transcription, but we have not studied this rigorously.

Gold: There does seem to be some disagreement on this point. The spinal nerve ligation (SNL) model is a model of nerve injury that has been used to determine whether nerve injury results in changes in Na$^+$ channel expression in 'uninjured' nerves. The sciatic nerve is comprised of axons arising from L4, L5 and L6 dorsal root ganglia. If you cut one or two of the spinal nerves, say L5 and L6, before they join the common nerve, the result is a pain syndrome with symptoms similar to those observed in humans following a partial nerve injury. The nice thing about this model is that it is a reproducible way of generating a partial nerve injury that allows one to study the relative contribution of injured and uninjured afferents to the pain behaviour.

Simon Tate and colleagues recently reported preliminary data indicating that there was no change of SNS ($Na_v1.8$) and SNS2 ($Na_v1.9$) immunoreactivity in the uninjured ganglia (Tate et al 2000). Porreca et al (1999) originally reported an increase in $Na_v1.8$ in the large cells of the uninjured ganglia, using a peroxidase method to detect $Na_v1.8$ immunoreactivity. I have collaborated with Porreca and colleagues in a follow-up study in which we were unable to detect a change in $Na_v1.8$ expression in the uninjured L4 ganglia using fluorescence immunohistochemistry. At the mRNA level Boucher et al (2000) recently reported an increase in $Na_v1.8$ expression in uninjured ganglia.

Waxman: There are contentions in the literature of relatively small but real changes in the adjacent ganglia; we haven't looked at this yet.

Gold: Do you look at all for rH1? This relates to compensatory change in the SNL model. Sandra Chaplan has recently reported preliminary data indicating that there is an increase in rH1 in the uninjured ganglia (Chaplan et al 2000).

Waxman: We have not looked at rH1 in the axotomy model. We do see rH1 in astrocytes (Black et al 1998), but the function of Na^+ channels in astrocytes remains a mystery to me.

Horn: I don't know whether you would call this an acquired channelopathy, but there was a recent paper from Reinhart Rüdel's group in MS patients and patients with Guillain–Barré syndrome, about an endogenous pentapeptide that has biophysical effects on inactivation (Brinkmeier et al 2000). Do you know what isoform of Na^+ channel they were looking at?

Waxman: No, and I don't think they knew, either. It is important for us all to remember that there is a long and turbulent history to the idea of blocking factors in the serum or spinal fluid in patients with MS, going back to the 1960s. There have been many contentions that there were blocking factors, and then many failures to reproduce these results. What was different about the paper that you are citing was the apparent specificity. They not only had a blocking factor, but also claimed to identify a specific pentapeptide, both in the spinal fluid of MS patients and at lower concentrations in normals. They talked about the possibility of what they called an endocaine, and observed a lidocaine-like effect. This observation still has not been reproduced in other laboratories, as far as I know. The jury is still out: it is potentially a very important and different aspect of channel involvement in inflammatory disorders of the nervous system, if it turns out to be true. If the phenomenon is real, it must be confirmed. I must emphasize, however, that more work needs to be done.

Horn: Have other laboratories failed to reproduce this, or do you just not know of labs that have tried this?

Waxman: A number of labs are working on this at the moment. Hopefully we will soon hear what they have found.

Horn: If it holds up, would you call that, by definition, an acquired channelopathy?

Waxman: I don't know what I'd call it, but it sure would be interesting.

Gold: Do you have data on the size distribution of neurons that show up-regulation of brain 3, and does that correlate with what you would expect, at least in terms of the source of ectopic activity?

Waxman: By *in situ* hybridization on dorsal root ganglia, 3 goes up in all classes of neurons. Ted Cummins' physiology was done on C type neurons. More recently we have seen that there is an acceleration of re-priming in larger neurons, but it is quantitatively different: it is a smaller shift than we see in the C-type neurons. In terms of repriming, it is not fair to extrapolate in a numerical way from the small cells to the bigger cells.

Baker: Is there any change in threshold? Seeing as you are losing two Na^+ channels and gaining one, what happens to the threshold in the C fibres?

Waxman: We have not yet looked at that.

Noebels: I was wondering about the apparent selectivity of the SNS up-regulation: was it everywhere, and not just Purkinje cells?

Waxman: We don't know the full answer. We have looked at two other areas in experimental autoimmune encephalitis (EAE) where neurons send axons through tracts that are demyelinated. We have looked at red nucleus neurons that send their axons down the spinal cord, and we have looked at sensorimotor cortex neurons. We do not see changes that are nearly as dramatic as what we see in the cerebellum. We are in the midst of doing a survey of the entire neuraxis.

Noebels: Was there an effect of plaques in the cerebellar white matter? Why would they turn on a new channel?

Waxman: The MS patients we have studied did have plaques in white matter of the cerebellum. It is not known whether, and how, that triggered an up-regulation of SNS. Is it related to inflammation per se? Is it related to demyelination? Is it related to change in impulse trafficking consequent to demyelination. We don't know. Years ago, Joe Black and I showed that contact between an axon and an oligodendrocyte has an effect on Na^+ channel expression in that axon. More recently Ben Barres and his colleagues have shown that soluble protein molecules that are produced by oligodendrocytes can have an effect on Na^+ channel expression in axons. There are a variety of possible ways by which this effect could be mediated; we are sorting through them.

Noebels: Do you think from the immunocytochemistry that tracking of this novel channel protein could be hung up in the soma, or is it being exported to the axon or dendritic compartments?

Waxman: What we see is immunolabelling in the cell body and the proximal part of the apical dendrite. There is a tiny bit in the axon, but not in the initial segment. With this antibody and this channel — and with many other antibodies and many channels — what we are seeing is largely an intracytoplasmic pool of channel. We need to be very circumspect about whether the channel is plugged into the membrane or not.

Segal: In regions of the surface of a neuron where one type of Na^+ channel is concentrated have you looked at the distribution of other types of Na^+ channels? Is a region with a lot of one type of Na^+ channel more or less likely to have other types of Na^+ channels, or is there no general pattern?

Waxman: That is an important question, which gets at the issue of how a cell builds or rebuilds an electrogenic membrane. Years ago, when I was a student here in London, we looked at the issue of patterns of myelination. Some myelinated fibres are built so that they conduct as rapidly as possible; others are built to act as delay lines. In each case, the pattern of myelination matches the functional needs. Myelin thickness and node distance are matched to functional requirements. It is as if the cells somehow monitor their own behaviour. In the 1970s John Moore and I did computer simulations: along peripheral myelinated fibres, the density of the Na^+ channels is at the point that maximizes conduction velocity. How does the cell know this? We know very little about the feedback loops. We don't know how cells monitor their own channel densities and distributions, and we don't yet know the range of compensatory changes that exist. What we see in knockouts suggests that there may be compensatory changes: when you knock out expression of one channel, the expression of other channels goes up. It is a very interesting phenomenon.

Raman: In the traces that showed persistent current on the 500 ms timescale, which channel carries the very slow residual current?

Waxman: That was in TTX, so we believe that is the NaN current.

Raman: Do those slowly inactivating currents inactivate much more rapidly or completely at more positive potentials?

Waxman: You only see those currents if you hold the cell at hyperpolarized potentials. There is a robust current of 10–11 nA if the cell is held at −120 mV.

Cummins: You have to treat the cells very gingerly to see this current. They do inactivate at the more extreme depolarized potentials, but it is nowhere near as fast as the TTX-sensitive current inactivates. There may be a difference in the inactivation gate or the docking site.

Raman: I am really interested in knowing whether, under normal conditions, those very slowly inactivating channels will inactivate significantly during a train of spikes. If so, to what extent can that so-called persistent current contribute to the depolarization in a series of repetitive action potentials?

Cummins: It has been quite difficult for us to do an experiment like this in a live cell, at least with the patch–clamp electrodes. The current runs down once you patch the cell. It is the hardest Na^+ current I have ever worked with; it disappears so quickly. This is why we did some of the computer modelling studies, to try to get an idea of what it might be doing. We need to go back and try to look and something like that under tighter conditions.

Raman: In the computer simulation, is the duration of the spike so brief that you don't get a significant inactivation of the current during the spike?

Cummins: On the basis of what we have seen in the patch–clamp studies, we think it would inactivate, because it does get to the higher potentials.

Raman: When $Na_v1.3$ was up-regulated, you showed that the time course of recovery went from 61 ms to 14 ms. At what potential was that measured?

Cummins: That was probably at −80 mV. We looked at a range of potentials, from −140 mV to −60 mV, and recovery was faster at all potentials in the injured neurons.

Raman: Does that correlate with the normal recovery of expressed channels? Specifically, do the slowly inactivating channels recover more slowly?

Cummins: We think that the slowly repriming current is generated by PN1. In human embryonic kidney (HEK293) cells PN1 does have very slow re-priming. The small neurons aren't that excitable, so you don't see the repetitive firing that is seen in a cerebellar neuron.

Waxman: In our heterologous expression studies, the type 3 channel certainly is rapidly repriming: it doesn't exactly match what we see after axotomy, so the expression of other channels may also be changing.

Crill: How fast does the change in expression of Na^+ channels occur with the salt loading?

Waxman: We don't know. We looked at three and four days. Certainly, since it is a transcriptional change, it is going to take hours to a day or so to see a functional change in terms of membranes.

Bean: Which channel do you think is increasing the ramp current in the salt-loading model?

Waxman: We think it is 6.

Bean: It was interesting that in the particular ramp currents you showed, it appeared that there was also a shift in voltage dependence towards more negative voltages. Is that something you see frequently?

Cummins: Those were cells that are isolated from older animals, and so the recordings were very touch and go. I wouldn't be confident saying that there was definitely a shift.

Raman: Would that increase in steady-state current remain if you ramped the voltage in the opposite direction?

Cummins: We didn't try this.

Strichartz: With respect to computer simulations, I am always concerned about the effect of temperature. We do most of our measurements around 20 °C for obvious reasons. In particular, the temperature dependence — the so-called Q_{10} — for inactivation is very large. One could find that a more rapid removal of inactivation, which would in principle support repetitive firing from kinetics measured at 20 °C, turns into a very rapid onset of inactivation at 37 °C. Thus rather than getting hyperexcitability, you get hypoexcitability. This is an explanation that has been used to describe some of the reasons for the temperature dependent conduction failure in MS. Although it is a challenge, it might be useful to do some of these simulations based on kinetics at temperatures that begin to get up towards physiological levels, in order to get a more physiological simulation.

Waxman: That is a good point. Like most labs we tend to patch at room temperature.

Meisler: The gene *SCN7A* (encoding the channel formerly known as $Na_v2.1$, now known as Na_x) was recently knocked out, and this affected the animal's salt intake (Watanabe et al 2000). Have you looked at that channel?

Waxman: We haven't looked at Na_x. Until that paper appeared, like everyone else we wondered what that channel did. It was a striking finding. However, that paper did not show that $Na_v2.1$ was functioning as a channel.

Strichartz: Relating observations that we make in easily accessible soma to what might be going on in axons troubles me, mostly because I can't see the connection. For example, in experiments applying TTX to functionally identified single axons *invivo*, in normal rats even before any neuropathy has been induced, we see very few impulses that are insensitive to high concentrations of TTX. They all seem to cluster down in the concentration range which would correlate to the range of TTX-sensitive Na^+ channels. I wonder how important for axonal hyper-excitability these TTX-resistant channels are. It could be that they are more an indication to lead us further when we are looking in the soma about some other phenomena. In the peripheral nerve experiments that were intra-axonal, did you look, for example, at TTX sensitivity?

Waxman: This was human tissue, so we didn't have much of a chance. In addition, the first spike is TTX sensitive, and it acts as a trigger for the delayed depolarization. You can abolish everything with TTX, but this does not show that the delayed depolarization is TTX sensitive. Unless one clamps, one cannot answer that question. There are questions of relative spatial distribution of the various channels. There is also the puzzling observation that the mid-point for the h_∞ curve, for at least a significant part of the TTX-sensitive current in DRG neurons, is −80 mV or so. Thus many of the channels would be expected to be inactivated at rest. We are also left with the puzzle of how reduction in density of certain channels, such as SNS and NaN, feeds in to what might become hyperexcitability in axotomized DRG neurons. Ted Cummins has made the intriguing suggestion that the loss of a persistent Na^+ current would be expected to lead to a hyperpolarizing shift in resting potential and that might relieve resting inactivation on other fast transient channels.

Baker: Regarding the function of the current attributed to NaN, one of its intriguing features is that it activates so slowly, and this might present us with a problem in trying to understand what it does. I have tried to understand something of what the late current in large DRG neurons does. This fast-activating late current has a low threshold and can amplify a rapidly rising applied current in the subthreshold potential range (Baker 2000). What happens afterwards is that fast K^+ channels come into play, obviating the amplification. In order for NaN to amplify depolarizing currents, it is going to have to do it in the face of fast activating K^+ channels. I think it has a job on its hands.

Bostock: There is evidence that TTX-resistant Na^+ channels are functionally important primarily at sensory nerve terminals. Brock et el (1998) found that TTX blocked conduction in axons innervating the guinea pig cornea, but did not block nociceptive impulses in the nerve terminals in response to natural stimulation. Similarly, Strassman & Raymond (1999) found evidence in recordings from nerve endings in mouse dura that TTX-resistant Na^+ channels are a feature of the receptive, rather than the conductive portion of slowly conducting sensory axons. At sensory nerve terminals, the rates of depolarization by natural stimulation can be low, and therefore the densities of fast-activating K^+ currents are also likely to be low.

Baker: In the SNS knockout there is clear evidence that SNS must be a part of the TTX-resistant current that is operating in the terminals (Akopian et al 1999). I am not sure that this answers my point. Because of its slow kinetics, NaN may have a job in significantly shifting membrane potential, and amplifying any impinging depolarization, because fast K^+ channels activate and minimize the change in membrane potential.

Horn: NaN may change the resting potential.

Baker: Absolutely.

Waxman: We viewed this in a simplistic sense as a conductance that modulates the standing resting potential. For me, one of the striking features of the computer simulations was that the depolarizing influence on resting potential is predicted to be non-trivial. It was a substantial depolarizing influence, about 15 mV. I agree that none of these channels operates in isolation. Even by looking at the ensemble of Na^+ channels, we are keeping blinders on and neglecting the cacophony of electrogenesis. There are some very fast K^+ channels and some

Ca^{2+} channels, and Ca^{2+}-activated K^+ channels: we have talked about just one set of players.

Spruston: What fraction of the total Na^+ current in these cells is mediated by SNS and NaN?

Waxman: In a cell at resting potential the amount of current flowing through NaN is small, but it is flowing through the cell at rest and so may have a disproportionately large effect because membrane conductance is low.

Spruston: If you give a step to 0 mV, activate a large fraction of all the Na^+ conductances in the cell, and then put on TTX, what fraction is left?

Cummins: That is a complicated question, because it depends on where you start from. It may be very different in the cell body than it is at the terminals. I don't know what the resting potential is out in the axon and at the nerve terminal: this may be where the important resting potential is. In the cell body, if I hold at very negative potentials, I will see roughly equal densities of current for TTX-sensitive and TTX-resistant channels. If I just look at channel densities, at the maximum channel availability the breakdown is about 50:50 in small neurons.

Bevan: If one looks across a spectrum of neurons and not just at small ones, even with cells with both it will vary significantly. Some cells have very little TTX-resistant current, and in other cells it dominates.

Strichartz: Those are the densitites of peak currents in response to voltage pulses, as opposed to currents that are activated and contribute to the regenerative current that is important in the rising phase of the action potential. The relative contributions can depend a lot on the slope and the actual rate of the rise. In this regard, the role of K^+ channels becomes increasingly critical: you can have a shunting outward current or leak conductance that are almost equivalent.

Spruston: The current clamp recordings from the $SNS^{-/-}$ mice imply that SNS contributes a dramatic amount of current during the rising phase of the action potential, because the amplitude is reduced dramatically. Is that a consistent observation? Also, the after-hyperpolarization is much more negative in the $SNS^{-/-}$ mice, implying that there is an actual contribution by SNS during the repolarizing phase of the action potential.

Waxman: We are in the midst of calculating the pNa:pK ratio at the peak of the spike in the wild-types and SNS nulls. We don't yet have all the data in, but it looks as if during the rise in the spike, the SNS channels contribute a substantial part of the Na^+ conductance.

Spruston: It seems to me that you are implying that there may also be up-regulation of a K^+ current in those animals.

Waxman: I don't know: what I know is that there are complex differences in the after-potentials.

Spruston: If it is up-regulation of the K^+ channels, that would be an odd way for a cell to respond to down-regulation and knockout of a Na^+ channel.

Gold: There were just two voltage traces illustrated and they were from different resting membrane potentials. These cells have a number of K^+ currents that are subject to steady-state inactivation where the steep part of the inactivation curve coincides with resting membrane potential. The result is that differences in available current may explain the apparent differences in after-hyperpolarization.

Waxman: From what we have seen in terms of Na^+ channels it is not surprising that K^+ channel expression is also quite dynamic. Jeff Kocsis and his colleagues have shown that following axotomy there is a down-regulation of some K^+ currents in DRG neurons. We have seen a down-regulation of some of the proteins.

Bostock: There is one important thing to bear in mind about the membrane potential of C fibres. In recordings from human nociceptors we have seen that it changes remarkably during ordinary activity, through activation of the Na^+ pump (Serra et al 1999). A nociceptor conducting at just 2 Hz will hyperpolarize progressively so that the latency may increase by 50% and the threshold goes up several times. We are talking about large changes in membrane potential by what seem to be low rates of activity. During burst activity the hyperpolarization can be enormous.

Waxman: We saw the same thing in the optic nerve. We did not include an electrogenic pump into our model, because it would have been one more unconstrained variable. But there is no question, if you activate the ATPase, it can cause hyperpolarization.

Bostock: On that matter, rates as low as one stimulus every 4 s will change the membrane potential, so that the after-potential goes from being depolarizing to hyperpolarizing. An axon that at rest is subexcitable after a nerve impulse, at very low rates of activation will become superexcitable after an impulse. This has great effects on its propensity for repetitive discharge. The membrane potential depends on Na^+ influx, but a very important effect is the indirect effect in activating the Na^+ pump.

References

Akopian AN, Souslova V, England S et al 1999 The tetrodotoxin-resistant sodium channel SNS has a specialized function in pain pathways. Nat Neurosci 2:541–548

Baker MD 2000 Selective block of late Na^+ current by local anaesthetics in rat large sensory neurones. Br J Pharmacol 129:1617–1626

Black JA, Dib-Hajj S, Cohen S, Hinson AW, Waxman SG 1998 Glial cells have heart: rH1 Na^+ channel mRNA and protein in spinal cord astrocytes. Glia 23:200–208

Boucher TJ, Okuse K, Bennett DL, Munson JB, Wood JN, McMahon SB 2000 Potent analgesic effects of GDNF in neuropathic pain states. Science 290:124–127

Brinkmeier H, Aulkemeyer P, Wollinsky KH, Rüdel R 2000 An endogenous pentapeptide acting as a sodium channel blocker in inflammatory autoimmune disorders of the central nervous system. Nat Med 6:808–811

Brock JA, McLachlan EM, Belmonte C 1998 Tetrodotoxin-resistant impulses in single nociceptor nerve terminals in guinea pig cornea. J Physiol 512:211–217

Chaplan SR, Calcutt NA, Higuera ES 2000 Presence of cardiac sodium channel mRNA in DRG and cord of experimental neuropathy models. Soc Neurosci Abstr 26:1111

Porreca F, Lai J, Bian D et al 1999 A comparison of the potential role of the terodotoxin-insensitive sodium channels, PN3/SNS and NaN/SNS2, in rat models of chronic pain. Proc Natl Acad Sci USA 96:7640–7644

Serra J, Campero M, Ochoa J, Bostock H 1999 Activity-dependent slowing of conduction differentiates functional subtypes of C fibres innervating human skin. J Physiol 515:799–811

Strassman AM, Raymond SA 1999 Electrophysiological evidence for tetrodotoxin-resistant sodium channels in slowly conducting dural sensory fibres. J Neurophysiol 81:413–424

Tate S, Decosterd I, Ji I et al 2000 Modulation of the TTX-R sensory neuron specific sodium channel SNS and SNS2 in diverse rat models of neuropathic pain. Soc Neurosci Abstr 26:939

Watanabe E, Fujikawa A, Matsunaga H et al 2000 Na_v2/NaG channel is involved in control of salt-intake behavior in the CNS. J Neurosci 20:7743–7751

Enhanced transmission of glutamate current flowing from the dendrite to the soma in rat neocortical layer 5 neurons

Wayne E. Crill, Peter C. Schwindt and John C. Oakley

Department of Physiology and Biophysics, University of Washington, 1959 NE Pacific Street, Box 357290, Seattle, WA 98195-7280, USA

Abstract. The presence of tetrodotoxin (TTX)-sensitive slowly-inactivating Na^+ channels in the dendrites of neocortical layer 5 neurons was tested by focal iontophoresis of glutamate on the dendrite while voltage clamping the soma and proximal dendrite. The glutamate-transmitted current was measured with the voltage clamp circuit. When the soma was depolarized the transmitted current increased indicating voltage-dependent properties in the dendrite. Over 50% of this increased voltage-dependence was blocked by TTX indicating a large portion of the enhanced dendro-somatic current was caused non-inactivating Na^+ channel inward rectification. The glutamate-transmitted current measured with a voltage clamp of the soma at firing level was equal to the effective glutamate measured during repetitive firing.

2002 Sodium channels and neuronal hyperexcitability. Wiley, Chichester (Novartis Foundation Symposium 241) p 61–71

The dendrites of central neurons receive thousands of synaptic inputs. The amount of this synaptic current transmitted to the spike-generating region, usually located in the proximal axon, determines the effectiveness of synaptic inputs. A major advance in understanding the functional properties of dendrites occurred with the quantitative analysis of passive dendritic filtering. The early investigators of dendritic cable properties also realized that the presence of voltage-dependent ion conductances in the dendrites could markedly affect their filtering properties (Jack et al 1975). For example, outward rectification in the dendrite would increase the electrical length, thereby increasing the attenuation of synaptic signals. On the other hand, inward rectification shortens the electrical length of the dendrites and would decrease the spatial attenuation of synaptic signals.

Several early studies using patch techniques have revealed voltage-dependent Na^+ channels in the apical dendrites of pyramidal shaped cells from the mammalian CNS (Magee & Johnston 1995, Spruston et al 1995, Stuart &

Sakmann 1994). Since the persistent Na^+ current in mammalian neurons is best explained by a modal change of inactivation time course, we hypothesized that the dendrites might also have a small but persistent voltage-dependent inward Na^+ current. When this inward rectification is activated it will increase the transmission of synaptic current to the spike-generating region in the proximal axon. Jack et al (1975) also recognized that inward rectification need not evoke an all-or-nothing response. Any inward rectification could decrease the spatial attenuation of a cable current. To test the hypothesis we elected to activate dendritic receptors by the focal iontophoresis of glutamate rather then by synaptic excitation. This approach was used because tetrodotoxin (TTX) must be used to establish the role of dendritic voltage-dependent Na^+ channels. TTX could also alter evoked test synaptic responses by changing synaptic input making interpretation of results difficult. The use of iontophoretically applied glutamate avoids this potential complication.

Slices 350–400 μm thick were taken from neocortex of Sprague–Dawley rats 21–35 days old using techniques described previously (Greene et al 1994). In some experiments 100 μM D-2-amino-5-phosphonopentoic acid (APV) or 1–10 μM MK-801 was added to the perfusate and $MnCl_2$ was substituted for $CaCl_2$. The apical dendrite or its branches were approached with an extracellular iontophoretic electrode containing 0.5–1 M glutamate. Locally evoked glutamate currents in the dendrites were recorded at the soma from an intracellular sharp or whole cell patch soma electrode in the voltage-clamp mode using an Axoclamp-2A amplifier. Thus the potential at the soma was controlled by the voltage clamp but voltage control of the dendrites decreased with distance from the soma. The responses were stable. The magnitude of the ligand-gated current changed when the iontophoretic electrode was moved only a few micrometers and often disappeared after moving it just 10 μm.

Figure 1A shows a cartoon of the experimental set-up. The neuron was voltage-clamped at resting potential (−76 mV in Fig. 1B). The dashed line indicates the one-second iontophoresis of glutamate 260 μm from the soma (Hu & Hvalby 1992). When the cell body was clamped at resting potential, the inward, glutamate-evoked current was not affected by the addition of TTX to the bathing solution. Voltage-clamping the soma to −61 mV increased the magnitude of the current flowing to the soma and a significant fraction of glutamate-evoked current was blocked by the TTX (Fig. 1C). The transmitted current was measured as the difference between the steady baseline current at the holding potential and the peak current observed during the iontophoresis (ΔI shown in Fig. 1B). Figure 1D is a plot of the transmitted current (ΔI) versus the clamped membrane potential, illustrating the graded, voltage-sensitive glutamate current blocked by TTX. Figure 1E shows the current–voltage relationship measured at soma holding potentials and illustrates that TTX-induced decrease in the glutamate current

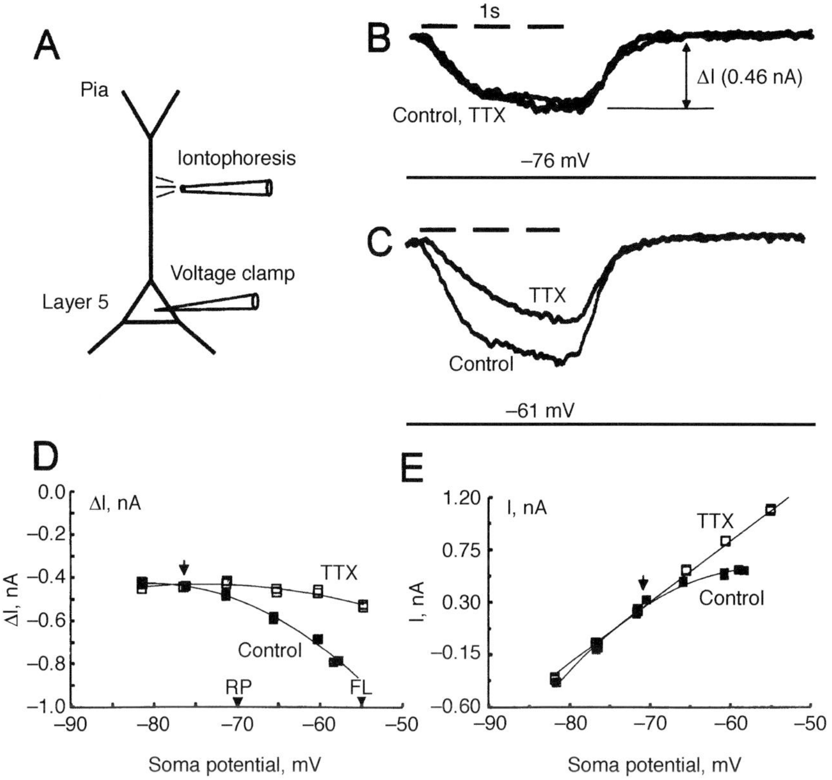

FIG. 1. Dendritic somatofugal glutamate-induced current is increased when soma and proximal dendrite are depolarized. (A) Experimental set-up. (B) Superimposed traces of transmitted current (ΔI) measured with somatic voltage clamp during one-second dendritic iontophoresis of glutamate. Voltage clamp holding potential at resting potential (−76 mV). (C) Superimposed traces, as in B, at voltage clamp holding potential of −61 mV before (Control) and after application of TTX. (D) Plot of transmitted current (ΔI) as function of soma voltage clamp holding potential. Arrowheads on abscissa mark resting potential (RP) and firing level (FL). Solid squares are Control and open squares after application of TTX. (E) Current–voltage relationship of baseline soma current before (Control) and after application of TTX. From Schwindt & Crill (1995).

(Fig. 1D) occurs at the potentials where the persistent TTX-sensitive current is activated. The test solutions contained APV to block voltage-dependent ligand gated current and 2 mM Cs^+ to block hyperpolarization activated current (I_H). Whereas the glutamate current increased with depolarization (Fig. 1D, dark squares), current evoked by iontophoresis applied to the soma decreased, as expected, when the soma was voltage-clamped to depolarized potentials, closer to the glutamate reversal potential. The somatic voltage-dependent channels cannot contribute to the amplitude because somatic potential is held constant by the voltage clamp.

These results depend upon the soma voltage clamp depolarizing the dendrite between the intracellular electrode at the soma and the iontophoretic site but this region is not isopotential with the soma because the dendrite is an extended cable. Since the transmitted current increases with soma depolarization there must be a voltage-dependent inward current summing with the glutamate current. Otherwise the transmitted current would either decrease with depolarization or show no change at all.

The TTX-sensitive portion of the transmitted current was prolonged indicating a persistent Na^+ current in the dendrites which 'amplifies' the transmitted ligand-gated current. The graded increase in the TTX-sensitive transmitted current with dendritic depolarization shown in Fig. 1D indicates that the transmitted current was not caused by a brief TTX-sensitive response evoking a prolonged all-or-nothing Ca^{2+}-mediated action potential. Furthermore, similar results were recorded when $MnCl_2$ replaced Ca^{2+} in the bathing solution.

Our observations demonstrate that ligand-gated depolarizing currents flowing to the soma are increased by non-inactivating or slowly inactivating dendritic Na^+ channels (Alzheimer et al 1993, Stafstrom et al 1985). Note that these voltage-dependent dendritic channels need not evoke all-or-nothing dendritic response. All that is required is for the dendritic current–voltage relation to show inward rectification. These biophysical properties of the dendrite effectively shorten the dendritic electrical length of the dendrite increasing the effectiveness of distal dendrites. If the filtering properties of dendrites were changed by synaptic activity the effectiveness of distal synapses would change.

NMDA receptors at the site of iontophoresis contributed to the amplified transmitted current reaching the soma since part of the voltage-dependent increase in the transmitted current was removed by treatment with APV or MK801.

One can ask whether the 'amplified' glutamate current or synaptic current occurs physiologically since the transmitted current was measured only when the voltage is held constant. During repetitive firing the somatic potential varies and there is good evidence for back-propagated action potentials from the soma into the apical dendrite. This normal physiological activity could affect the current transmitted to the soma. To examine these effects we compared the transmitted current during soma voltage clamp to that measured during repetitive firing.

To estimate effective synaptic current during repetitive firing we used a method developed for spinal motor neurons (Powers et al 1992). Synaptic current and soma-injected current add algebraically. For example, the addition of a steady synaptic current to an injected soma current causes a parallel shift in the frequency–current (f–I) curve. This implies that the faster firing rate caused by synaptic current could not be distinguished from firing evoked by current injected in the soma. Powers et al (1992) have shown that this effective current computed from the f–I curve equals the synaptic current arriving at the soma. This occurs because the spike-generating region is down stream from the soma. Injected current into layer 5 pyramidal neurons causes a parallel shift in the f–I curve (Schwindt & Crill 1996).

In our experiments the f–I relationship of the layer 5 pyramidal cell was measured by injecting constant currents of various strengths into the neuron. We then measured the steady firing rate evoked by dendritic glutamate iontophoresis, which allowed calculation of the effective current depolarizing the soma. The effective glutamate current reaching the soma during repetitive firing was compared with the glutamate current flowing into the soma during voltage clamp (Fig. 1). If the iontophoresed current did not evoke repetitive firing it was added to current injected into the soma to allow calculation of the effective glutamate current by the repetitive firing method.

An example of the results from these experiments in a neocortical layer 5 pyramidal neuron is shown in Fig. 2. Figure 2A shows the response of the neuron to injected soma current alone and Fig. 2B shows the increased firing when added to dendritic glutamate iontophoresis. The instantaneous firing rate is nearly constant (Fig. 2C). The parallel shift of the f–I relationship during iontophoresis is shown in Fig. 2D. In the same cell during soma voltage clamp (Fig. 2E) the effective current during different levels of soma depolarization (voltage clamp) are plotted (Fig. 2F).

Is the effective glutamate current during repetitive firing similar to the glutamate current measured during voltage clamp of the soma near resting potential when dendritic voltage-dependent conductances are not activated, or is it similar to the transmitted glutamate current near threshold where the attenuation caused by the dendritic cable properties is less? To answer this question the effective glutamate current measured during repetitive firing was compared to the transmitted glutamate current at resting potential and at firing level (Fig. 3). The transmitted glutamate current measured at the firing level corresponds best with the transmitted current measured by repetitive firing.

The repetitive firing experiments show that injected current into the soma causes a parallel shift in the f–I relationship, which supports the concept of a downstream location of the spike-generating site. Secondly these experiments show that the glutamate-injected current evokes repetitive firing in the physiological range. Finally these experiments reveal that the transmitted glutamate current during physiologic repetitive firing is equal to the 'amplified' glutamate current during voltage clamp.

These experiments directly reveal the change in effectiveness of ligand-gated current caused by voltage-dependent dendritic channels. Activation of inward

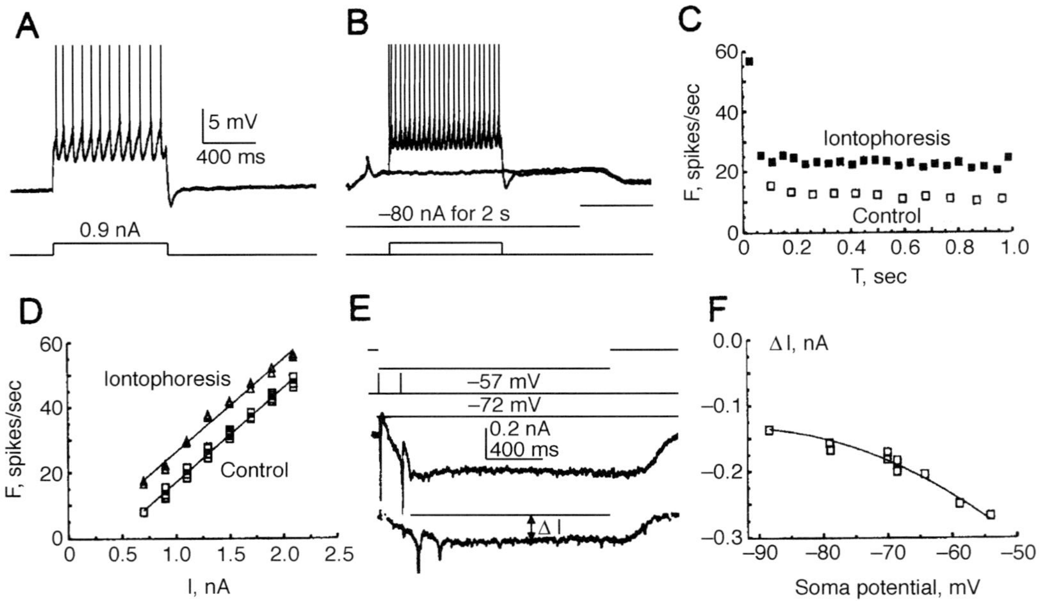

FIG. 2. Measurement of dendritic glutamate current during repetitive firing and soma voltage clamp. (A) Repetitive firing evoked 0.9 nA current injected into the soma. (B) Superimposed traces dendritic iontophoresis (middle trace) induced depolarization and repetitive firing evoked by same current as in A during iontophoresis. (C) Instantaneous firing during applied soma current (open squares) and during dendritic iontophoresis (closed squares). (D) Iontophoresis of glutamate on the dendrite causes a parallel vertical shift in the frequency (F)–current (I) relationship. (E) Transmitted current (ΔI in bottom traces) measured at two voltage clamp holding potentials of the soma. Top trace shows glutamate iontophoresis. (F) Transmitted current (ΔI) as function of soma voltage clamp holding potential. From Schwindt & Crill (1996).

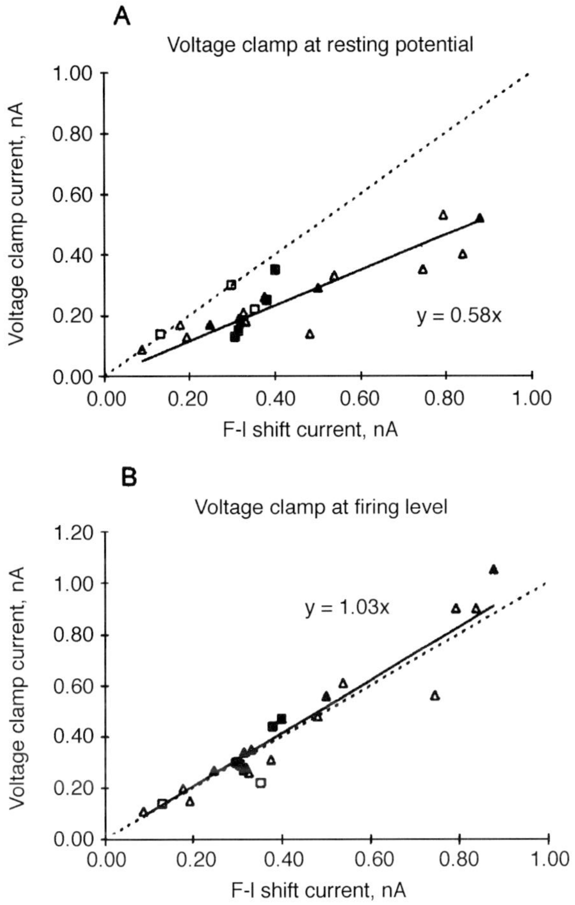

FIG. 3. Comparison of transmitted current measured with voltage clamp (ordinate) with current measured by repetitive firing method (abcissa). Each symbol is a different cell. For perfect agreement the regression line for the points (solid line) would lie on the dashed line. From Schwindt & Crill (1996).

rectifying dendritic current, including a slowly inactivating Na^+ current, shortens the electrical length of the dendrite thereby increasing the effectiveness of synaptic input to a particular dendritic site. For example, the uniform excitation of the dendrites by synaptic input will progressively increase the strength of the input and this will occur without the generation of dendritic spikes. Persistent Na^+ currents, Ca^{2+} currents and NMDA synaptic currents all contribute to this increase in synaptic effectiveness. Numerous types of voltage-dependent ion channels have been identified in the dendrites. At least in the rat neocortical slice depolarization causes a net inward rectification rather than voltage-dependent outward currents, which would have the opposite effect. The balance of inward and outward rectification could be modulated by synaptic transmitters and alter the effectiveness of synapses at a specific geometric distance without a change in transmitter release or transmitter receptor properties.

References

Alzheimer C, Schwindt PC, Crill WE 1993 Modal gating of Na^+ channels as a mechanism of persistent Na^+ currents in pyramidal neurons from rat and cat sensorimotor cortex. J Neurosci 13:660–673

Greene C, Schwindt PC, Crill WE 1994 Properties and ionic mechanisms of metabotropic glutamate receptor-mediated slow afterdepolarization in neocortical neurons. J Neurophysiol 72:693–704

Hu GY, Hvalby O 1992 Glutamate-induced action potentials are preceded by regenerative prepotentials in rat hippocampal pyramidal cells in vitro. Exp Brain Res 88:485–494

Jack JJB, Noble D, Tsien RW 1975 Electric current flow in excitable cells. Clarendon, Oxford

Magee JC, Johnston D 1995 Synaptic activation of voltage-gated channels in the dendrites of hippocampal pyramidal neurons. Science 268:301–304

Powers RK, Robinson FR, Konodi MA, Binder MD 1992 Effective synaptic current can be estimated from measurements of neuronal discharge. J Neurophysiol 68:964–968

Schwindt PC, Crill WE 1995 Amplification of synaptic current by persistent sodium conductance in apical dendrite of neocortical neurons. J Neurophysiol 74:2220–2224

Schwindt PC, Crill WE 1996 Equivalence of amplified current flowing from dendrite to soma measured by alteration of repetitive firing and by voltage clamp in layer 5 pyramidal neurons. J Neurophysiol 76:3731–3739

Spruston N, Schiller Y, Stuart G, Sakmann B 1995 Activity-dependent action potential invasion and calcium influx into hippocampal CA1 dendrites. Science 268:297–300

Stafstrom CE, Schwindt PC, Chubb MC, Crill WE 1985 Properties of persistent sodium conductance and calcium conductance of layer V neurons from cat sensorimotor cortex in vitro. J Neurophysiol 53:153–170

Stuart GJ, Sakmann B 1994 Active propagation of somatic action potentials into neocortical pyramidal cell dendrites. Nature 367:69–72

DISCUSSION

Spruston: When you do the experiment with the glutamate iontophoresis, the reason that you get additional TTX-sensitive current coming in must be because you have voltage escape in the dendrites. If you had perfect voltage clamp, you wouldn't see this.

Crill: That's exactly right.

Spruston: Given that this is the case, how do you know that any small depolarization produced by Na^+ channels doesn't also activate Ca^{2+} channels? The extra current that you measure could be a mixture of Na^+ and other voltage-activated currents that are activated by the subsequent voltage change.

Crill: I refer to the response produced by TTX. There is a component of the amplification that is mediated by NMDA channels, and Ca^{2+} blockers will knock out part of this. The bulk of the amplification is TTX-sensitive.

Spruston: How relevant is the depolarization induced by glutamate iontophoresis to those that actually occur with synaptic inputs? How much of the depolarization is occurring in the dendrites and what duration does this have when you do your glutamate iontophoresis, compared to what happens with synaptic stimulation?

Crill: The prolonged plateaus may be relatively unphysiological, although in awake animals there are periods of prolonged firing, which may or may not be due to underlying plateau responses in the dendrites. We have been able to evoke these types of responses by synchronous stimulation of synaptic input to the dendrites. This is still not truly physiological, but it is closer than pouring on glutamate. The reason we picked the plateau is because that was something we can control experimentally and where we could get steady-state firing, so we could make some quantitative measurements. I didn't point it out, but in the changes in the plateau that you see from evoking at multiple sites, you see similar types of changes occurring in the more transient Ca^{2+} spikes. We would like to believe that what we say about the plateaus applies also to the more frequently occurring Ca^{2+} spikes.

Spruston: Going back to my first question, the main issue I was getting at is that there is still some debate over the question of how common persistent Na^+ current is in dendrites of these and other neurons. It is clear that there are Na^+ currents. The question is, how much persistent Na^+ current influx can you get? Given that you showed that there are plateau potentials mediated by Ca^{2+} currents, it is possible that there is an initial depolarization mediated by Na^+ currents which is transient, which subsequently activates a Ca^{2+} current that is more sustained, producing an unclamped voltage change up in the dendrites. So the TTX-sensitive current may in fact not be a persistent Na^+ current, but could be a transient Na^+ current giving rise to a depolarization and activating a persistent Ca^{2+} current. Do you have any evidence that would rule out that scenario?

Crill: We can still see the amplification in the presence of Ca^{2+} channel blockers, so that does happen. I would not say it is only Na^+ channels, though. I am sure there are some Ca^{2+} channels: in some imaging studies that I didn't show, there is significant Ca^{2+}.

Strichartz: What is the evidence that the passive properties for these very small processes with large surface:volume ratios are changed as a result of Na^+ accumulation, affecting for example delayed rectifier K^+ channels or Ca^{2+}-activated K^+ channels? In most of the modelling, you assumed that the passive properties were invariant: you just added an active process on top of that.

Crill: This is Julian Jack's modelling, not ours. They just assumed that the dendrite has a given current:voltage relationship.

Strichartz: Have you looked for changes in passive properties that would persist after these active properties are over, but during a period when intracellular ions might still be elevated?

Crill: We haven't done those specific experiments, but it is worth emphasizing that since there is clearly a diversity of channels in the dendrites, there may be significant variation in the expression and effectiveness of these channels over time. This can markedly shift the electrical closeness of input to the soma and therefore the effectiveness of a given synaptic input. How this happens is a subject for future experiments.

Noebels: I was struck by the resemblance of the linear jump shift transition to a plateau that you saw, and the data that Hodgkin, Stein and others have described for axons (Jack et al 1975). Can intense or persistent depolarization cause not only a spread of the initiation site to properties of proximal dendrites, but also beyond the axon to possibly even the first node? This would result in spike initiation at the first node that could backfire and contribute in some way to the f–I parameters that you see.

Crill: We haven't asked that question, but I think the evidence against it is that when you inject current into the soma, this isn't seen at all. We can measure a linear f–I relationship if we inject the current into the soma; it is only when we depolarize the dendrites that we get that saturation of frequency of firing.

Noebels: In some way, perhaps the dendrites are becoming more axon-like, in the sense that many axons have that same f–I behaviour.

Waxman: Some years ago, Mahlon Kriebel, Mike Bennett and I studied oculomotor neurons in fish. They have spike trigger zones in the cell body and in the dendrites, but when they fire at high frequency, you can see the spike initiation zone migrating from one site to the other. It is not a fixed patch of membrane as you are implying.

Bean: I am curious about the possible involvement of inactivating K^+ channels in the jump. Another possible mechanism would be some sort of A-type K^+ channels that inactivated with voltage. These could also give you regenerative depolarization. I don't remember whether A-type K^+ channels are present in these dendrites or not.

Spruston: There are two recent papers both showing that these neurons do not have nearly as much IA as hippocampal pyramidal neurons do, for example (Bekkers 2000, Korngreen & Sakmann 2000). They definitely don't have the strong gradient that is present in CA1 neurons.

Crill: The type of response we get is definitely localized. The Ca^{2+} goes up wherever you are ionophoresing. Exactly what keeps it localized is unknown. We have done some modelling to try to see whether we could reproduce the behaviour, but this has been unsuccessful. However, it clearly exits experimentally.

Spruston: In CA1 neurons, something analogous happens. If you have a train of spikes that are paired with current injection into the dendrites, you can get Ca^{2+} spikes. Here, it seems that a crucial role is played by the D-type K^+ currents. These D-type K^+ currents seem to limit the generation of Ca^{2+} spikes near the soma, and these currents are activated by the action potentials themselves. So if you block either the D-type K^+ currents or the fast Na^+ spikes, this dramatically lowers the threshold for Ca^{2+} spikes in the soma of CA1 pyramidal neurons. It is possible that something similar is happening in the layer 5 neurons (Golding et al 1999). Matthew Larkum has published a very nice paper in which he showed that if you pair back-propagating action potentials with synaptic input in the layer 5 neurons, then you get generation of Ca^{2+} spikes in the dendrites of the layer 5 cells (Larkum et al 1999). Probably what is also happening with the jump is that when you get high frequency spike trains paired with glutamate ionophoresis, it generates a plateau potential in the dendrite.

Crill: I want to make it clear that although I referred to these as Ca^{2+} spikes, there are probably other inward currents involved. In every cell that we have examined we could block the plateau of response under Ca^{2+} spikes with cadmium, but in a few cells we could also block it with NMDA blockers or TTX. There seems to be some variation, but primarily the charge carrier seems to be Ca^{2+}.

References

Bekkers JM 2000 Distribution and activation of voltage-gated potassium channels in cell-attached and outside-out patches from large layer 5 cortical pyramidal neurons of the rat. J Physiol 525:611–620

Golding NL, Jung HY, Mickus T, Spruston N 1999 Dendritic calcium spike initiation and repolarization are controlled by distinct potassium channel subtypes in CA1 pyramidal neurons. J Neurosci 19:8789–8798

Jack JJB, Noble D, Tsien RW 1975 Electric current flow in excitable cells. Clarendon, Oxford

Korngreen A, Sakmann B 2000 Voltage gated K^+ channels in layer 5 neocortical pyramidal neurones from young rats: subtypes and gradients. J Physiol 525:621–639

Larkum ME, Zhu JJ, Sakmann B 1999 A new cellular mechanism for coupling inputs arriving at different cortical layers. Nature 398:338–341

Mutations of voltage-gated sodium channels in movement disorders and epilepsy

Miriam H. Meisler, Jennifer A. Kearney, Leslie K. Sprunger, Bryan T. MacDonald, David A. Buchner and Andrew Escayg

Department of Human Genetics, University of Michigan, 4909 Buhl Box 0618, Ann Arbor, MI 48109–0618, USA

Abstract. Spontaneous and induced mutations of neuronal Na^+ channels in human patients and mutant mice result in a broad range of neurological disease. Epilepsy, a disorder of neuronal hyperexcitability, has been associated with delayed inactivation of *SCN2A* in mice, and with altered kinetics of *SCN1A* in human patients. Movement disorders including tremor, ataxia, dystonia and paralysis have been observed in mice with mutations of *SCN8A*. Electrophysiological recordings from neurons isolated from mice with mutations in individual channels reveal the contributions of each channel to *in vivo* firing patterns. In addition to monogenic disease, Na^+ channel mutations are likely to contribute to polygenic disease susceptibility and to normal variation in neuronal function. Advances in molecular methods coupled with genomic sequences from the Human Genome Project will permit identification of many new patient mutations and generation of animal models to dissect their physiological and cellular consequences.

2002 Sodium channels and neuronal hyperexcitability. Wiley, Chichester (Novartis Foundation Symposium 241) p 72–86

Other papers in this volume address the biophysical properties of individual voltage-gated Na^+ channels. *In vivo* Na^+ currents in neurons are the result of the combined activity of multiple Na^+ channels, as well as K^+ channels and other functionally interacting proteins. The contributions of individual channels *in vivo* can be inferred by analysis of mice with mutations in a single Na^+ channel. The neurological abnormalities in the mutant animal provide insight into the role of the mutated channel in genetic disorders, and electrophysiological recordings from brain sections or single cells can detect the effects of the mutations on sodium currents in specific types of neurons. In this chapter we describe neurological and cellular abnormalities caused by mutations of three neuronal Na^+ channel genes, *SCN1A*, *SCN2A* and *SCN8A*.

The GAL879-881Q3 mutation in Na^+ channel *SCN2A* causes seizures in mice

Muscle and cardiac disease result from mutations that delay the inactivation rate of Na^+ channels expressed in those tissues, but similar mutations in the neuronal Na^+ channels have not been described (Bulman 1997). To determine the effect of a delayed inactivation mutation in neurons, we introduced a mutation into the *SCN2A* cDNA encoding the Na^+ channel $Na_v1.2$, or brain type 2. We used the neuron-specific enolase promoter to direct expression of the mutant cDNA in transgenic mice. The mutant, GAL879-881QQQ in the cytoplasmic S4–S5 linker of domain 2, has a slightly delayed rate of inactivation and a small increase in persistent current when assayed in *Xenopus* oocytes (Fig. 1A). The effect on mice expressing the transgene was dramatic (Kearney et al 2001). When the amount of mutant channel was equal to 20% of the wild-type channel, the mice exhibited severe seizures and died by two months of age. When the amount of mutant channel was only 2% of the wild-type channel, as in line Q54, the mice survived for 6–8 months, and had seizures beginning at 3–4 months of age and increasing in frequency and severity. In addition to seizures, the Q54 mice exhibited frozen postures and repetitive behaviours such as grooming or swimming movements (Fig. 2). Tonic/clonic seizures in Q54 mice were accompanied by EEG patterns indicative of hippocampal seizures; these occurred between three and 40 times per day. Na^+ currents were recorded from hippocampal CA1 neurons by Ted Cummins, who observed persistent currents similar to those generated by the same mutation in the *Xenopus* oocyte assay (Fig. 1B). A few weeks after the initiation of seizures, neuronal cell loss and gliosis was evident throughout the hippocampus (Kearney et al 2001). The seizure disorder in Q54 mice thus begins with persistent currents in hippocampal neurons, followed typically by initiation

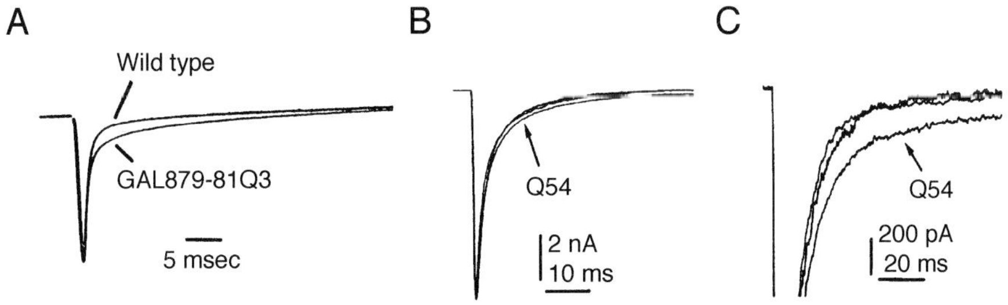

FIG. 1. The GAL879-881QQQ mutation in the S4/S5 linker of domain 2 of SCN2A results in increased persistent current. (A) Kinetics of mutant and wild-type channel in *Xenopus* oocytes. Normalized current traces during a depolarization to -5 mV. (B) Voltage-dependent Na^+ currents in hippocampal CA1 neurons from Q54 mice. (C) 10× magnification of B.

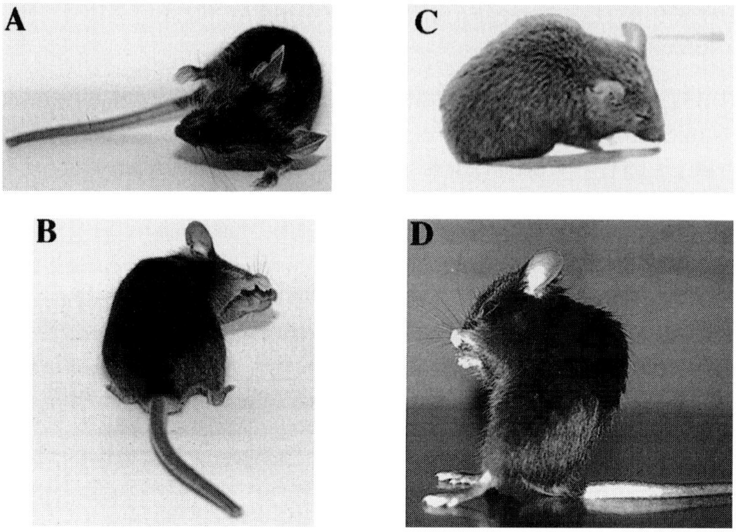

FIG. 2. Appearance of GAL879-881QQQ transgenic mice during seizures. (A) and (B) Focal motor seizures with tonic deviation of the head and body and clonic forelimb movement. (C) This frozen posture was accompanied by spike wave discharges and rhythmic EEG activity. (D) An unusual frozen posture that was maintained for 5 min. Reprinted from Kearney et al (2001) with permission.

of seizures at 2 months, cell loss at 4 months, and death at 6–8 months. Persistent current thus may lead directly to hyperexcitability of the CA1 neurons in Q54 mice.

This mutant line will be a useful model for testing the effectiveness of pharmacological intervention in preventing cell loss, an important goal of epilepsy treatment in human patients. Since other Na^+ channels are also expressed in hippocampal neurons, it seems likely that delayed inactivation of *SCN1A*, *SCN3A* or *SCN8A* could also result in seizures.

Mutations of *SCN1A* result in generalized epilepsy in human patients

In 1999, a locus for the human epilepsy syndrome GEFS+ (generalized epilepsy with febrile seizures plus) was mapped to the region of chromosome 2q24 that includes the Na^+ channel genes *SCN1A*, *SCN2A* and *SCN3A* (Baulac et al 1999, Moulard et al 1999). Individuals with GEFS+ have a heterogeneous phenotype, with febrile seizures in childhood progressing to several types of generalized seizures in adults. To determine whether the underlying cause of these seizures was similar to that in the Q54 mice, we tested each exon of *SCN1A* and *SCN2A* by conformation-sensitive gel electrophoresis (Escayg et al 2000).

Mutations of *SCN1A* were detected in both families (Fig. 3). The mutations introduce substitutions for phylogenetically conserved residues in S4 domains of the channel: a threonine to methionine mutation at the cytoplasmic end of D2S4 (T875M) and an arginine to histidine mutation in the R5 residue of D4S4 (R1648H). Al Goldin has investigated the kinetics of the mutant channels in the *Xenopus* oocyte assay and identified abnormalities in both mutants (Spampanato et al 2001). These are the first mutations in a human neuronal Na^+ channel α subunit to be associated with neurological disease, and, like the Q54 mice, demonstrate an association between Na^+ channels and neuronal hyperexcitability.

Mutations of *SCN8A* cause movement disorders in the mouse

SCN8A is another major Na^+ channel that is widely distributed in the CNS and PNS. SCN8A appears to be the only Na^+ channel located in the nodes of Ranvier in adults, and is also found in cell bodies, dendrites, and presynaptic and postsynaptic membranes (Caldwell et al 2000, Schaller & Caldwell 2000, Krzemien et al 2000). We have been studying four independent spontaneous

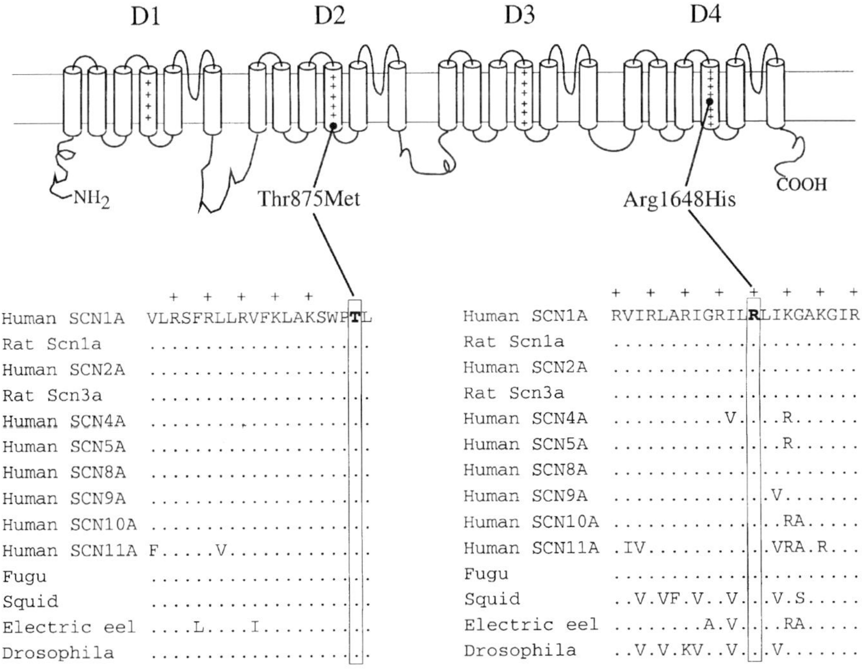

FIG. 3. Two *SCN1A* mutations in patients with familial epilepsy. Thr875 and Arg1648 in D2S4 and D4S4 are conserved in vertebrate and invertebrate channels.

mutations in *SCN8A* in the mouse, two null alleles, one hypomorph with 2% of the normal level of expression, and one amino acid substitution, A1071T (reviewed in Meisler et al 1997, 2001). Each mutation results in a distinct neurological disorder at the whole animal level and specific changes in Na^+ currents at the cellular level (Table 1). The phenotypes of the mutant mice have provided insight into the biological role and medical implications of changes in this channel.

Complete loss of SCN8A is lethal

The nodal localization of SCN8A would predict that mice cannot survive without this channel. In fact, null mutants lacking this protein are paralysed and do not survive beyond one month. The timing of death may be explained by an apparent developmental switch in the type of Na^+ channel located at the node. Between 2 and 3 weeks of age, Na^+ channels SCN1A ($Na_v1.1$) and SCN2A ($Na_v1.2$) can be detected at the nodes with specific antisera. In the third week of life, during the period of myelination, these channels appear to be replaced by SCN8A ($Na_v1.6$), which is the only channel detected in nodes in adult mice (Caldwell et al 2000). Consistent with this time course, the null mutants lacking SCN8A (*med* and *med*^{tg}) can walk at 2 weeks of age but become paralysed by 3 weeks of age. Secondary effects of the null mutations include sprouting of nerve terminals at the neuromuscular junction (Duchen & Stefani 1971) and atrophy of the hind limb muscles.

The SCN8A mutation A1071T in jolting mice results in tremor and ataxia

Transmission at the neuromuscular junction is normal in jolting mice, indicating that this mutation does not impair function at the nodes of Ranvier (Harris & Pollard 1986). Introduction of the jolting mutation into the *SCN2A* and *SCN8A* cDNAs by site-directed mutation in Al Goldin's laboratory resulted in a 10 mV rightward shift in the voltage dependence of activation assayed in *Xenopus* oocytes (Kohrman et al 1996a, Smith & Goldin 1999). Recordings from cerebellar Purkinje cells isolated from jolting mice by Indira Raman and Bruce Bean (Raman et al 1997) detected major changes in firing patterns. Purkinje cells produce a pattern of complex spikes in response to a single current injection, but fewer spikes were generated in cells with the jolting mutation (Fig. 4). The resurgent current and persistent current which are also characteristic of Purkinje cells were almost completely lost in the mutant (Fig. 4). These data demonstrate that the SCN8A channel is essential for the characteristic firing properties of cerebellar Purkinje cells. Ataxia in jolting mice may be secondary to these cerebellar effects. Spontaneous firing of cartwheel cells of the dorsal cochlear nucleus, which are ontologically related to Purkinje cells, is also reduced in *Scn8a* mutant mice

TABLE 1 Four independent mutations in the mouse *Scn8a* gene encoding $Na_v1.6$

Mutant	*Allele*	*Effect on protein*	*Molecular mutation*	*Neurological phenotype*	*Physiological phenotypes*	*Cellular phenotypes*
med	$Scn8a^{med}$	Null	Spontaneous line1 insertion	Paralysis	Muscle atrophy, slowed nerve conduction, neurotransmitter release at NMJ function	Sprouting of motor neurons muscle, atrophy
TgNA4Bs	$Scn8a^{tg}$	Null	Non-targeted transgene insertion	Paralysis	Same as above	Purkinje cells, motor neurons, cortical pyramidal cells
med^{J}	$Scn8a^{medJ}$	Hypomorph (2% level of transcript)	Splice site deletion	Weakness, dystonia		P-cells, DCN, motor neurons
jolting	$Scn8a^{jo}$	Amino acid substitution	Nucleotide substitution	Tremor, ataxia		Purkinje cells but not motor neurons

P-cell, cerebellar Purkinje neurons; DCN, dorsal cochlear nucleus; NMJ, neuromuscular junction.

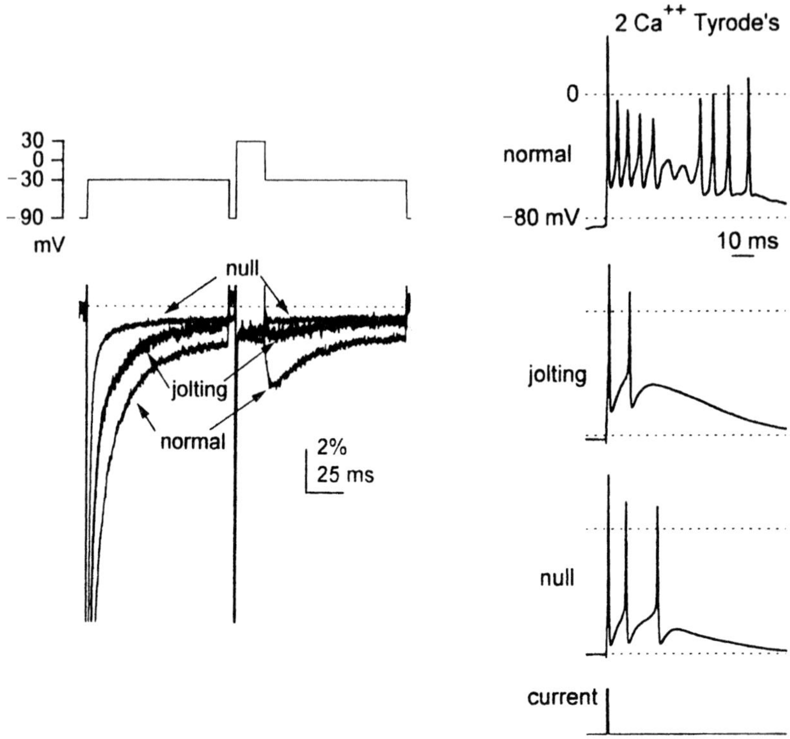

FIG. 4. Two characteristic currents of cerebellar Purkinje cells are reduced in mice with mutation in *Scn8a*. Left, resurgent current; right, persistent current. Reprinted from Raman et al (1997) with permission.

(Chen et al 1999). It is interesting that the amino acid substitution A1071T has a major effect on Purkinje neurons but no evident effect on the function of motor neurons.

A modifier gene determines survival of mice deficient in SCN8A

The *med*^{J} mutation is a hypomorphic allele of *SCN8A*. Due to a splice site mutation, the amount of correctly spliced *Scn8a* transcript in *med*^{J} mice is only 2% of the normal level (Kohrman et al 1996b, Sprunger et al 1999). The neurological phenotype of homozygous *med*^{J} mice can be changed by crossing the mutation onto different background strains. On most strains, *med*^{J} mice have muscle weakness, persistent tremor and dystonic postures with twisted trunk and limbs that are maintained for several minutes at a time (Fig. 5A). In one strain, C57BL/6, *med*^{J} homzygotes have the more severe phenotype of progressive

FIG. 5. A modifier gene alters the clinical consequences of low levels of *Scn8a* in homozygous *med*^{J} mice. The 8 month old homozyogte on strain C3H exhibits muscle weakness and dystonic postures. The 3 week old animal on strain C57BL/6J is paralysed. The difference is caused by the *Scnm1* modifier locus on chromosome 3.

paralysis and juvenile death, similar to the null mutants (Fig. 5B). This phenomenon is reminiscent of the clinical differences between affected individuals within human pedigrees who share the same mutation but may differ at other 'modifier' genes that influence the phenotype. Genetic crosses demonstrated that the difference between C57BL/6J and the other strains is due to allelic differences in one major modifier gene, designated *SCNM1* (sodium channel modifier 1) (Sprunger et al 1999). We are attempting to clone the modifier gene, with the goal of identifying a novel protein with a role in Na^+ channel processing or function.

Conclusions and future prospects

These initial studies of mice with spontaneous or induced mutations in neuronal Na^+ channel genes suggest that the orthologous human channels are candidate genes for many human disorders. Electrophysiological characterization of neurons isolated from the mutant mice has provided new information about the contributions of these channels *in vivo*. Future identification of human disease mutations will be facilitated by the genomic sequence from the Human Genome Project, which permits the design of specific primers used for mutation detection. Mouse models carrying the specific molecular defects found in human patients can be generated using conventional trangenes or targeting in embryonic stem (ES) cells. We anticipate an explosion in the identification and modelling of neurological disorders resulting from Na^+ channel mutations during the next few years. The current status of *in vivo* Na^+ channel mutations is summarized in Table 2.

TABLE 2 Physiological consequences of Na^+ channel mutations *in vivo*

Gene	*Protein*	*Tissue*	*Species*	*Major phenotype (mutation)*
SCN1A	$Na_v1.1$	Neuron	Human	GEFS+ epilepsy (missense)
SCN2A	$Na_v1.2$	Neuron	Mouse	Brain stem apoptosis (null), epilepsy (Q54)
SCN4A	$Na_v1.4$	Muscle	Human	Episodic paralysis (missense)
SCN5A	$Na_v1.5$	Muscle	Human	Long QT syndrome (missense)
SCN7A	Na_x	Many	Mouse	Abnormal salt intake (null)
SCN8A	$Na_v1.6$	Neuron	Mouse	Paralysis (null), ataxia (missense), dystonia (low expression)
SCN10A	$Na_v1.8$	Neuron	Mouse	Pain sensitivity (null)

In addition to monogenic diseases like GEFS+, polymorphic variants of voltage gated Na^+ channels are likely to contribute to susceptibility to common polygenic disorders including psychiatric diseases. In surveys of human populations, we have detected coding variants in the neuronal Na^+ channels, some rare and some common. It will soon be feasible to design a 're-sequencing chip' that would detect all single nucleotide changes in human ion channel genes by hybridization of PCR products against arrayed oligonucleotides. In addition to the medical benefits of identifying disease mutations for diagnosis and development of treatments, biophysical analysis of mutations with impaired *in vivo* function will expand our understanding of structure/function relationships. Together these two approaches, the detection of human mutations and the generation of mutant mice, will provide a wealth of research material that will lead to better understanding Na^+ channel function.

Acknowledgements

Our work on the Na^+ channel genes has been supported by the National Institute of General Medical Sciences (GM24872), the National Institute of Neurological Diseases and Stroke (NS34509), the March of Dimes, and the American Epilepsy Foundation with support from UCB Pharma, Inc.

Note added in proof

During the past year, several mutations responsible for epilepsy have been identified in SCN1A and SCN2A (reviewed in Meisler 2001).

References

Baulac S, Gourfinkel-An I, Picard F et al 1999 A second locus for familial generalized epilepsy with febrile seizures plus maps to chromosome 2q21-q33. Am J Hum Genet 65:1078–1085

Bulman DE 1997 Phenotype variation and newcomers in ion channel disorders. Hum Mol Genet 6:1679–1685

Caldwell JH, Schaller KL, Lasher RS, Peles E, Levinson SR 2000 Sodium channel Na(v)1.6 is localized at nodes of Ranvier, dendrites, and synapses. Proc Natl Acad Sci USA 97: 5616–5620

Chen K, Sprunger LK, Meisler MH, Waller HJ, Godfrey DA 1999 Reduced spontaneous activity in the dorsal cochlear nucleus of Scn8a mutant mice. Brain Res 847:85–89

Duchen LW, Stefani E 1971 Electrophysiological studies of neuromuscular transmission in hereditary 'motor end-plate disease' of the mouse. J Physiol 212:535–548

Escayg A, MacDonald BT, Meisler MH et al 2000 Mutations of SCN1A, encoding a neuronal sodium channel, in two families with GEFS+2. Nat Genet 24:343–345

Harris JB, Pollard SL 1986 Neuromuscular transmission in the murine mutants 'motor end-plate disease' and 'jolting'. J Neurol Sci 76:239–253

Kearney JA, Plummer NW, Smith MR et al 2001 A gain-of-function mutation in the sodium channel gene SCN2A results in seizures and behavioral abnormalities. Neuroscience 102: 307–317

Kohrman DC, Smith MR, Goldin AL, Harris JB, Meisler MH 1996a A missense mutation in the sodium channel gene Scn8a is responsible for cerebellar ataxia in the mouse mutant jolting. J Neurosci 16:5993–5999

Kohrman DC, Harris JB, Meisler MH 1996b Mutation detection in the med and med^J alleles of the sodium channel Scn8a: unusual splicing due to a minor class AT-AC intron. J Biol Chem 271:17576–17581

Krzemien DM, Schaller KL, Levinson SR, Caldwell JH 2000 Immunolocalization of sodium channel isoform NaCh6 in the nervous system. J Comp Neurol 420:70–83

Meisler M 2001 Genetics of epilepsy. Annu Rev Genet 35

Meisler MH, Sprunger LK, Plummer NW, Escayg A, Jones JM 1997 Ion channnel mutations in mouse models of inherited neurological disease. Ann Med 29:569–574

Meisler MH, Kearney J, Escayg A, MacDonald BT, Sprunger LK 2001 Sodium channels and neurological disease: insights from Scn8a mutations in the mouse. Neuroscientist 7:136–145

Moulard B, Guipponi M, Chaigne D, Mouthon D, Buresi C, Malafosse A 1999 Identification of a new locus for generalized epilepsy with febrile seizures plus (GEFS+) on chromosome 2q24-q33). Am J Hum Genet 65:1396–1400

Raman IM, Sprunger LK, Meisler MH, Bean BP 1997 Altered subthreshold sodium currents and disrupted firing patterns in Purkinje neurons of Scn8a mutant mice. Neuron 19:881–891

Schaller KL, Caldwell JH 2000 Developmental and regional expression of sodium channel isoform NaCh6 in the rat central nervous system. J Comp Neurol 420:84–97

Smith MR, Goldin AL 1999 A mutation that causes ataxia shifts the voltage-dependence of the Scn8a sodium channel. Neuroreport 10:3027–3031

Spampanato J, Escayg A, Meisler MH, Goldin AL 2001 Functional effects of two voltage-gated sodium channel mutations that cause generalized epilepsy with febrile seizures plus type 2. J Neurosci 21: 7481–7490

Sprunger LK, Escayg A, Tallaksen-Greene S, Albin RL, Meisler MH 1999 Dystonia associated with mutation of the neuronal sodium channel Scn8a and identification of the modifier locus Scnm1 on mouse chromosome 3. Hum Mol Genet 8:471–479

DISCUSSION

Ptacek: How narrow is the region containing the modifier gene? Is it within that 2 Mb?

Meisler: Yes, we are now down to about 1 Mb. I think we can rescue it, because it is a lethal phenotype.

Noebels: Even though it only survives to be 3 weeks old, isn't the null $Na_v1.6$ mouse living evidence that $Na_v1.6$ can't be the only nodal Na^+ channel?

Meisler: We have always wondered why these animals get around much better at 2 weeks of age and then decline by 3 weeks of age. It now seems that between 2 and 3 weeks postnatal in the normal animal there is a replacement of $Na_v1.2$ at the nodes by $Na_v1.6$. So the animals survive until the time that $Na_v1.6$ becomes the major nodal channel at around 3 weeks of age.

Noebels: It would be interesting to expand the expression studies to look at other channels, to see whether there is some attempt at plasticity in these mice that lack the $Na_v1.6$ subunit. The null $Na_v1.6$ mice must still have nerve conduction if they live to be 3 weeks old.

Meisler: I think this is the normal developmental pattern. Other channels like $Na_v1.2$ are present at the nodes in the younger animals, followed by replacement with $Na_v1.6$ between 2 and 3 weeks of age. This doesn't happen in the null $Na_v1.6$ mice and therefore they die.

Ptacek: In the *SCN1A* R1648 mutation, it looked like there was a hyperpolarizing shift in the voltage dependence of inactivation.

Goldin: There is an 8 mV negative shift when the α subunit is expressed alone; when it co-expressed with the β subunit, there is no shift.

Ptacek: Then it looked like there was an even smaller rightwards shift of the activation curve.

Goldin: There was no significant shift. The line is a little bit to the right, but it is not significantly different. This was with or without the $\beta1$ subunit.

Horn: If there is very fast recovery from inactivation and there was no shift, this means that the entry into the inactivated state must also be very fast.

Goldin: It should mean that, but so far we haven't looked at this carefully enough. With the two-electrode clamp there were no gross differences in the kinetics of entry into the inactivated state.

Ptacek: Miriam Meisler, early on you said that knockouts won't have phenotypes. What did you mean by this?

Meisler: What I meant to say was that I don't think we are going to find null alleles in human populations, because the effects would be so severe.

Ptacek: From the knockout experiments in mice, it seems the heterozygotes are usually normal.

Meisler: Yes, it appears that 50% of the Na^+ channels are sufficient for normal function.

Ptacek: Homozygous would almost certainly be lethal. However, *SCN4A*, for example, may be a candidate for the cause of a muscle disorder such as congenital fibre type disproportion, where there are fewer fast twitch fibres. One hypothesis might be that loss of a single allele causing 50% protein loss might lead to a problem of muscle contraction only during maximal effort.

Meisler: It doesn't seem to be the case in the mice we have looked at: the heterozygotes appear normal.

Ptacek: It would also be interesting to look in the mice at seizure thresholds with proconvulsants. Again, you can imagine a heterozygous loss of function could give a predisposing factor in the human population that you would never see in homozygosity.

Meisler: That would be interesting to look at. My guess would be that you wouldn't see it, because of the cytoplasmic pool of excess channels. It doesn't appear that channel production is rate-limiting for these channels. Haplo-insufficiency in general only affects fewer than 5% of genes. These are usually structural genes such as collagen. In most cases, 50% is adequate for normal function. However, a loss of function allele of SCN1A was recently found in patients with Severe Myoclonic Epilepsy of Infancy (Claes et al 2001).

Ptacek: Again, when you start moving from Mendelian genetics into susceptbility or predisposition, these would be really attractive alleles to use as candidates.

Bean: Have you looked at pain sensing in these channel mutants?

Meisler: No. That would be very interesting.

Segal: Have you looked at the ability to treat the seizures in these mutants with Na^+ channel acting drugs? This could give us clues about whether the seizures are due to the increase in Na^+ channels in the present time or whether they are due to some past effect such as kindling that is giving some other type of seizure.

Meisler: That would be one good use for the SCN2A mutant. We haven't done this yet.

Strichartz: One of the problems with that approach is that you can have a basic pathology whose origins are quite different than the Na^+ channels, but the hyperactivity manifests through the Na^+ channels. Suppressing that hyperactivity does not inform you about the basic mechanism of the lesion.

Meisler: What if you are starting with a mouse in which you know the basic lesion is in a Na^+ channel?

Strichartz: Then I don't understand the reason for using the pharmacological approach.

Segal: A failure to treat the seizures using Na^+ channel anticonvulsants would suggest that the seizures were being expressed in a secondary fashion, for example resulting from kindling or neuronal death.

Noebels: It is curious that the seizures don't start until the mice are about three months old, but the abnormal currents can be seen much earlier. Is it really because of the delayed current (which you have obviously changed), or does this cause other downstream changes in the nervous system which ultimately make it pathological? We don't know.

Goldin: The seizures are clearly progressive. Something is getting worse with time, and it is not the channel properties. Cells are dying, and as cells are killed off in the hippocampus, things are getting worse.

Noebels: Is it actually neuronal cell loss that causes seizures in the epileptic brain?

Meisler: We can see neuronal cell loss.

Waxman: Why are the cells dying? Are they bursting and burning themselves out, or is there a persistent Na^+ current letting a Na^+ influx drive reverse Na^+/Ca^{2+} exchange? Or is it that these animals are becoming transiently hypoxic during seizures and sustaining damage because of that?

Meisler: The damage seems to be localized.

Cummins: It could be hypoxia: the hippocampus is one of the regions that does get hit fairly early by hypoxia.

Strichartz: Is it clear that the cells that express the mutation are the ones that are dying? Are they releasing high concentrations of glutamate locally and killing the cells that are responding?

Meisler: We expect that all of the cells are expressing this construct. It is the NSE promoter for the neural-specific enolase gene. However, we haven't done histology to demonstrate pan-neuronal expression, because the level of expression is so low it is hard to detect.

Ptacek: It would be interesting to make that mutant under an inducible promoter, because then just by flipping it on at different times or in different locations you could address the question of temporal and spatial effects.

Goldin: We could turn it on at a later time and avoid the issue of developmental changes.

Ptacek: If you were to turn it on at 8 weeks and see seizures immediately, this would suggest that it is not a degenerative disease.

Goldin: It would also allow us to determine how long it takes. NSE is turning on during the first two or three weeks after birth, and the phenotype is starting after that. It is difficult to know exactly what is going on.

Segal: It would also be helpful if you could target the Na^+ channel abnormality to certain neuronal populations.

Meisler: We certainly could. Which would you suggest?

Segal: The hippocampus would be of interest because of its sensitivity to epileptiform activity and neuronal death. Perhaps if you lack the Na^+ channel abnormality in the hippocampus, then you won't go on to develop a lot of the secondary types of seizures. It would be interesting to turn on or off the Na^+ channel mutations in CA1 or CA3 pyramidal neurons.

Cummins: I have a general comment about the neuronal disease mutations. Louis Ptacek identified one of the hyperkalaemic periodic paralysis mutations (T704M) back in the early 1990s and the first three studies that looked at the T704M mutation found rather subtle defects (Cannon & Strittmatter 1993, Cummins et al 1993, Yang et al 1994). But the T704M mutation actually has a major effect on slow inactivation (Cummins & Sigworth 1996), and none of us caught that in the first three studies. It could be that the neuronal mutations that Miriam Meisler has described affect some property that hasn't been looked at yet. There are a lot of different properties to look at, and therefore it is going to be complex to work out exactly how a given mutation causes a disease.

Goldin: The human $Na_v1.1$ domain 2 mutant that Miriam found in GEFS+ doesn't show any obvious changes. We are sure the changes are there; we just haven't looked at the right property yet.

Horn: There are a finite number of properties of the Na^+ channels (at least those that we know about at the moment), which wouldn't be too hard to put into a screen. You can see activation, fast inactivation, kinetics and steady-state of gating and slow inactivation. Those protocols are not so difficult to do on an individual cell. The slow inactivation ones are tedious to do, however!

Strichartz: I would argue that current clamp, following excitability and action potentials driven by short and long current injections, is a more sensitive assay than voltage clamp. As few as 2% of the channels can be non-inactivating, and give remarkable changes in excitability patterns. It would be very hard to pick this up under voltage clamp.

Horn: But the results would be uninterpretable.

Strichartz: You pick up the abnormality, and then you go hunting for those 2%. What fraction of the Na^+ channels in the hyperkalaemic mutant are actually non-inactivating? It is just 3–8%. Had you not seen a behavioural mutant, you wouldn't have picked it up: it would have been written off as biological variation.

Cummins: Even the 3–8% estimate for non-inactivating channels is still considered controversial. Steve Cannon has proposed that mechanism for hyperkalaemic periodic paralysis, but it has not been proven in muscle cells. 10 years following Louis Ptacek's discovery of the first hyperkalaemic periodic paralysis mutation, we are still not sure what the underlying mechanism is. I would argue that it is not due to a non-inactivating current, whereas Steve Cannon might argue that it is. We need to get these mutants back into the native cells and do a current clamp, which we can't do in oocytes or HEK293 cells.

Meisler: This is where the mutant mice will be useful, as a source of neurons for electrophysiology. Also, I think there will be mutations in the loops and in the C-terminus and N-terminus which may not affect the kinetics but do change cell biology. If you have a candidate human mutation, the only way to determine the functional consequence may be to introduce the mutation into the mouse.

Noebels: If there are modifier genes then you are still never sure.

Meisler: If you get an abnormal phenotype in the mouse you can be sure that the mutation is responsible. If you do not see any abnormal behaviour in the mouse, it would still be possible that the mutation had a negative effect in the human cells.

Ptacek: Are you going to be satisfied with the transgenic? To really recapitulate the *in vivo* situation in humans you would have to use homologous recombination to introduce the mutation into the native gene.

Meisler: Introduction of the mutation into the endogenous gene by targeting in ES cells does produce the best model of a human genetic disease. However, the cost, time and effort involved in ES cell targeting means that only a few mutations can be studied by that method. For mutations with a dominant mode of inheritance in humans, transgenic mice can provide useful and relevant models, much more quickly. It is possible to select transgenic lines in which the level of expression of the transgene is comparable to that of the endogenous gene. The main limitation of transgenes driven by promoters from other genes is that the foreign promoter does not precisely recapitulate the temporal and spatial pattern of expression of the endogenous gene. The use of BAC clones that contain the gene and its regulatory elements to create transgenic lines circumvents this problem. The mutation is introduced into a BAC clone by mutagenesis in bacterial culture, and the BAC DNA is then microinjected into fertilized mouse eggs. Expression of such BAC transgenes is independent of insertion site, and recapitulates the normal expression pattern. This new generation of transgenic mice provides a compromise between the speed of transgenics and the biological relevance of targeted alleles.

References

Cannon SC, Strittmatter SM 1993 Functional expression of sodium channel mutations identified in families with periodic paralysis. Neuron 10:317–326

Claes L, Del-Favero J, Ceulemans B, Lagae L, Van Broeckhoven C, De Jonghe P 2001 De novo mutations in the sodium-channel gene SCN1A cause severe myoclonic epilepsy of infancy. Am J Hum Genet 68:1327–1332

Cummins TR, Sigworth FJ 1996 Impaired slow inactivation in mutant sodium channels. Biophys J 71:227–236

Cummins TR, Zhou J, Sigworth FJ et al 1993 Functional consequences of a Na^+ channel mutation causing hyperkalemic periodic paralysis. Neuron 10:667–678

Yang N, Ji S, Zhou M et al 1994 Sodium channel mutations in paramyotonia congenita exhibit similar biophysical phenotypes in vitro. Proc Natl Acad Sci USA 91:12785–12789

Channelopathies: episodic disorders of the nervous system

Louis Ptacek

Howard Hughes Medical Institute, Departments of Neurology and Human Genetics, University of Utah School of Medicine, Eccles Institute of Human Genetics, Building 533, Room 4425, Salt Lake City, UT 84112, USA

Abstract. The field of channelopathies is a newly recognized group of disorders named after the site of their molecular defects — voltage- and ligand-gated ion channels. While voltage-gated ion channel mutants have been recognized for some time in organisms such as *Drosophila*, the first channelopathy in humans was reported within the last decade. The recognition of this group of disorders began with the definition of the molecular basis of a group of unusual muscle disorders called the periodic paralysis and non-dystrophic myotonias. Interestingly, this group of muscle disorders share some interesting phenotypic features with a number of seemingly disparate human diseases that involve not only skeletal muscle, but also brain and heart. Some similarities that exist among these different disorders include their episodic nature, similarities with regard to factors that precipitate attacks, therapeutic agents which can help to treat or prevent attacks, and in some cases, a degenerative component that arises in addition to the episodic attacks. The study of these diseases, along with the recognition of common clinical and pathophysiological themes among these disorders has led to tremendous growth in our understanding of these diseases and the hope of developing better therapies.

2002 Sodium channels and neuronal hyperexcitability. Wiley, Chichester (Novartis Foundation Symposium 241) p 87–108

Ion channel structure

Voltage-gated ion channels are proteins critical for establishment of the resting membrane potential in muscle and the ability of these membranes to generate action potentials. Voltage-gated opening of Na^+ channels results in the genesis of 'all-or-none' action potentials. These channels close over the course of a few milliseconds. A simultaneous effect of depolarization of the membrane (albeit on a slower time scale) is the opening of voltage-gated K^+ channels. Along with the inactivation of the voltage-gated Na^+ channels, the movement of positive K^+ ions out of the cell through these K^+ channels leads to a relatively rapid repolarization of the muscle membranes. Cl^- channels are responsible for a majority of the polarity

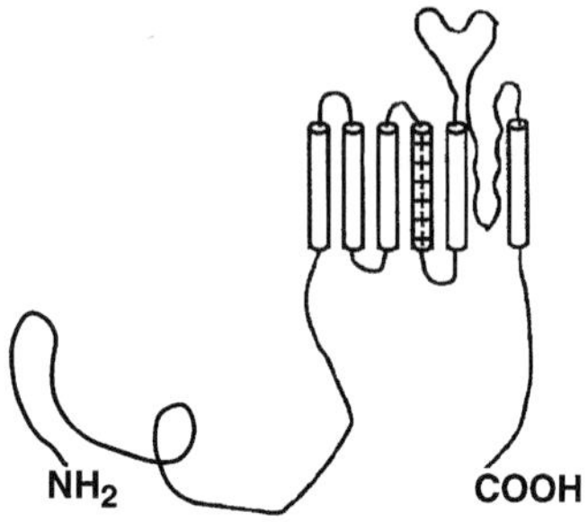

FIG. 1. Voltage-gated K^+ channel α subunit. The N-terminus and the C-terminus are intracellular. There are six transmembrane segments. The S4 segments contain positively charged arginine and lysine residues at every third position and are important in voltage sensing. Putative membrane-spanning S5 and S6 segments and the P region linking the two segments line the channel pore. Four subunits come together to form a functional K^+ channel.

of resting membranes. The voltage-gated L-type Ca^{2+} channel in skeletal muscle, also known as the dihydropyridine receptor because dihydropyridines block this channel, allows conductance of Ca^{2+} into the cell. It is likely that this Ca^{2+} is important through signalling pathways to effect other downstream changes of the muscle membrane but the Ca^{2+} conductance itself is not directly important to the depolarization of the muscle membrane. Interestingly, this voltage-gated Ca^{2+} channel serves as a voltage sensor for excitation–contraction coupling. Depolarization of the muscle membrane leads to changes in the Ca^{2+} channel protein, which in turn interacts with the ryanodine receptor to open slow release Ca^{2+} channels in the sarcoplasmic reticulum (SR). It is these slow release channels and the SR Ca^{2+} stores that lead to elevations in cytosolic Ca^{2+} that result in the ultimate contraction of muscle.

K^+ channels are membrane-bound tetrameric protein complexes. Each subunit is a polypeptide with six putative transmembrane segments (S1–6) (Fig. 1). Four identical subunits may associate to form a homomeric K^+ channel. Alternatively, different subunits may assemble to form a heteromeric K^+ channel. Mutagenesis studies have identified several critical functional domains in K^+ channel. The S4 segments with positively charged arginine and lysine residues at every third position appear important in voltage sensing. The P region linking the putative membrane-spanning S5 and S6 contains the K channel signature sequence that is highly conserved and critical for K selectivity of the channel pore.

The voltage dependent Na^+ and Ca^{2+} channels are a large group of homologous genes that are also homologous to the voltage-gated K^+ channel genes. Unlike the K^+ channel proteins where four subunits must come together to form a homo- or hetero-tetrameric functional channel, Na^+ and Ca^{2+} channel α subunits have

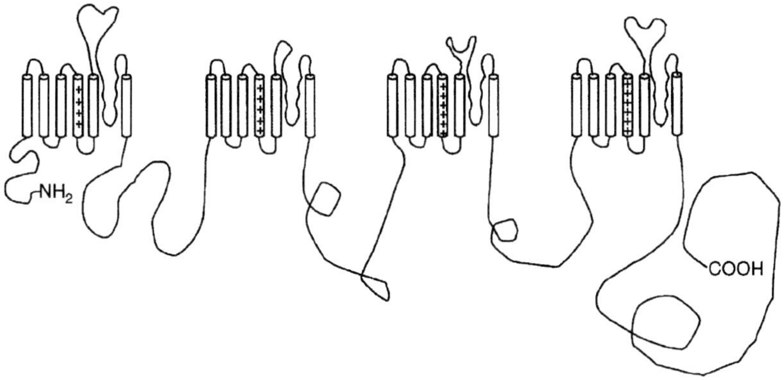

FIG. 2. Diagram of an α subunit of a Na^+ channel or α1 subunit of a Ca^{2+} channel. There are four homologous domains (internally homologous and homologous to the voltage-gated K^+ channel), each with six transmembrane segments.

evolved to include in a single transcript four domains (I–IV) in tandem, each with six transmembrane segments (S1–6) homologous to voltage-gated K^+ channel genes (Fig. 2). It is thought that this phenomenon resulted from the duplication of a progenitor 'K^+-like' channel that duplicated and then re-duplicated in the genome. The large, pore-forming α subunit alone is sufficient for cation permeability, pharmacologic specificity, gating and voltage sensitivity. The auxiliary subunits modulate channel biophysical properties and biosynthesis. Voltage-dependent Na^+ channels are composed of a pore-forming, voltage-sensing α subunit and a transmembrane β subunit. Voltage-dependent Ca^{2+} channels are heteromeric complexes composed of a pore-forming α1 subunit, a disulfide-linked membrane-anchored extracellular α2-δ subunit, an intracellular β subunit, and in muscle a γ subunit.

Voltage-gated Cl^- channels have been discovered more recently and much less is known about their structure. These proteins have 13 hydrophobic segments but more recent biochemical and immunological data suggest that not all of these traverse the membrane. It appears that voltage-gated Cl^- channels form multimeric functional units but the exact stoichiometry of these channels is not entirely clear. There are no recognized auxiliary subunits of the voltage-gated Cl^- channels at this time.

Periodic paralysis and non-dystrophic myotonia

Clinical manifestations

Periodic paralyses and nondystrophic myotonias include a number of distinct clinical entities as well as some intermediate forms of the various disorders. Myotonia congentia is a group of muscle disorders named for the prominent

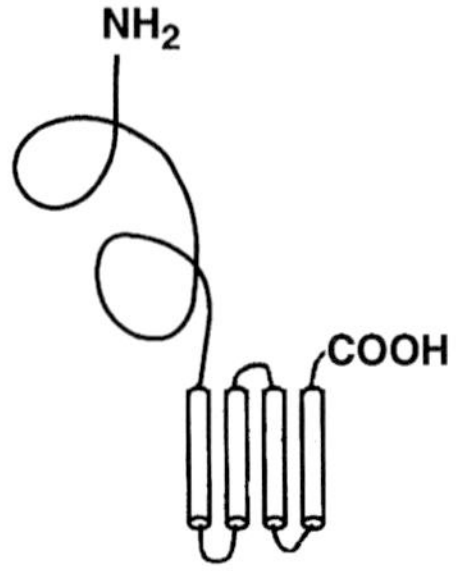

FIG. 3. Diagram of a subunit of nicotinic acetylcoline receptor channel or inhibitory glycine receptor channel. There are four membrane-spanning segments. These proteins are not homologous to voltage-gated channels.

muscle hyperexcitability or myotonia that is seen in these patients. This myotonia is classical myotonia with the phenomenon of 'warm up'. These patients experience extreme muscle stiffness due to delayed relaxation from repetitive electrical activities in muscle, but this myotonia subsides as their muscles warm up with use. Onset of symptoms is generally in childhood through early adult life. These patients often have hypertrophy of their muscles and a Herculian appearance as a result of their myotonia. Two distinct forms of myotonia congenita are recognized. The first, named for Julius Thompson who described the disease, is an autosomal dominant form of myotonia congenita (George et al 1994). These patients do not develop degeneration of their muscles even after years of having the disease. An autosomal recessive form of myotonia congenita was described by Becker (1971). These patients have myotonia with 'warm-up' phenomenon, may have transient bouts of weakness after periods of disuse, and sometimes develop myopathy as part of their disease.

Hyperkalaemic periodic paralysis is a disorder with myotonia like that seen in the above described disorders. These patients can also have a transition of their muscle membrane hyperexcitability to inexcitability in the form of episodic weakness. This weakness may be so dense as to cause a transient flaccid quadraparesis. However, the disorder does not affect the diaphragm and patients are therefore able to continue breathing. First described in 1951 by Tyler (Tyler et al 1951), this disorder is named because of the ability to precipitate attacks in patients by administrating a sufficient dose of oral K^+. During spontaneous attacks patients may have elevated K^+, although this is frequently in the normal range. Attacks of weakness can be precipitated by foods high in K^+, rest after vigorous exercise, and with stress and fatigue. The disease is transmitted as an autosomal dominant trait although sporadic cases are sometimes encountered. Percussion and action myotonia are frequently elicited clinically and prominent myotonia can be noted on an electromyographic exam. Patients benefit dramatically from treatment with carbonic anhydrase inhibitors.

Paramyotonia congenita is yet another disorder in which myotonia is present. Although the myotonia is somewhat different as it does not show the classical warm-up phenomenon, but rather is paradoxical. That is, patients frequently have worsening of their myotonia with repeated muscle action. This can be most prominently seen in the obicularis occuli muscles when patients forcefully close their eyes repeatedly and, upon such a manoeuvre, the myotonia of these muscles becomes increasingly severe to the point where patients might have difficulty opening their eyes altogether. Of interest, this disorder is a temperature-sensitive mutant of humans. With cooling of their muscles, these patients have worsening of their myotonia and then transition of the hyperexcitability into paralysis. This can be measured quantitatively using electrodiagnostic manoeuvres (Jackson et al 1994). Paramyotonia congenita is transmitted as an autosomal dominant trait, and like hyperkalaemic periodic paralysis, these patients have worsening of their symptoms with stress, fatigue, and rest after vigorous exercise. Patients may be hypo-, normo- or hyperkalaemic during attacks. The classical paramyotonia congenita patients are generally hypokalaemic during attacks; those with hyperkalaemia seem to represent a clinical entity somewhere in the spectrum between classical paramyotonia congenita and hyperkalaemic periodic paralysis. Like the hyperkalaemic periodic paralysis patients, these patients benefit dramatically from treatment with carbonic anhydrase inhibitors.

K^+-activated myotonia is a disorder where patients clinically appear to have myotonia congenita. But their myotonia fluctuates, worsens when K^+ is administered, and improves with carbonic anhydrase inhibitors (Trudell et al 1987). Interestingly, these patients do not develop attacks of weakness. This disorder is transmitted as an autosomal dominant trait.

Hypokalaemic periodic paralysis is a disorder of episodic weakness in which a myotonia is not seen. These patients are generally hypokalaemic during an attack and attacks can be precipitated by lowering K^+ with administration of glucose and insulin. Furthermore, these attacks can be precipitated by stress, fatigue and rest after vigorous exercise. Dietary precipitants include high carbohydrate meals and salt load. Hyperkalaemic periodic paralysis is transmitted as an autosomal dominant trait although frequent sporadic cases can be seen. Patients K^+ levels during hypokalaemia generally remain above 2 mM. K^+ levels below 2 mM during an attack of weakness in a sporadic case raises the possibility of thyrotoxic hypokalaemic periodic paralysis, a non-Mendelian form of the disorder which is seen only during periods of thyrotoxicosis (Ptacek 1998). The episodic weakness in patients with hyperkalaemic periodic paralysis, paramyotonia congenita and hypokalaemic periodic paralysis all benefit from treatment with the carbonic anhydrase inhibitors diamox and daranide.

Genetics

Linkage analysis in large pedigrees with hyperkalaemic periodic paralysis established that the gene for this disorder resided on chromosome 17q (Fontaine et al 1990, Ptacek et al 1991b). Mapping data showed that the paramyotonia congenita and the K^+-activated myotonia phenotypes also mapped to the same locus (Ptacek et al 1991c, 1992a). Subsequently, a Na^+ channel gene in this region of chromosome 17q was cloned and characterized and shown to be the site of mutations in hyperkalaemic periodic paralysis (Ptacek et al 1991a, Rojas et al 1991), paramyotonia congenita (McClatchey et al 1992, Ptacek et al 1992b), and K^+-activated myotonia (Ptacek et al 1994a). The data supporting that this Na^+ channel gene, *SCN4A* was the disease-causing gene included the following: (1) mutations segregated with the phenotype; (2) they involved highly conserved amino acid residues; (3) these mutations were not found in control individuals; and (4) some of them occurred as *de novo* mutations in patients with sporadic disease. A large number of Na^+ channel mutations have been identified in patients with hyperkalaemic periodic paralysis, paramyotonia congenita and K^+ aggravated myotonia.

Similar molecular approaches led to mapping of the hypokalaemic periodic paralysis locus to chromosome 1q (Fontaine et al 1994, Ptacek et al 1994b). Subsequently, a voltage-gated Ca^{2+} channel at this chromosome 1 locus was cloned. This Ca^{2+} channel gene encodes an L-type Ca^{2+} channel and three patient-specific mutations have been identified (Ptacek et al 1994b). The mutations occur at highly conserved residues in the S4 segments of either domain 2 and 4. The domain 4 mutations occur at an analogous position to the most common paramyotonia congenita mutations in the homologous *SCN4A* Na^+ channel (Ptacek et al 1994a).

Genetic linkage analysis in families segregating alleles for both recessive and dominant forms of myotonia congenita showed that both of these phenotypes were linked to a locus on chromosome 7q (Abdalla et al 1992, Koch et al 1992). Subsequent work has identified a voltage-gated Cl^- channel at this locus, *CLCN1*, to be the site of defects in myotonia congenita (Abdalla et al 1992, Koch et al 1992). A long list of mutations in this channel gene have since been shown to cause dominant and recessive forms of myotonia congenita (reviewed in Jen & Ptacek 2000). Interestingly, some of these mutations are recognized to cause both dominant and recessive forms (Zhang et al 1996). While it is likely that polymorphisms in the Cl^- channel gene itself may modulate the effect of such mutations, no data is available as yet to substantiate this hypothesis. Many of these Cl^- channel mutants have been expressed *in vitro* (either in oocytes or by transfection of mammalian cells growing in culture) and the physiologic abnormalities in myotonia congenita are being characterized. Discussion of the functional consequences is beyond the scope of this chapter but these are leading to new insights to these disorders.

Episodic ataxia

Clinical manifestations

Episodic ataxia is a rare inherited syndrome of intermittent ataxia of early onset with no known inborn errors of metabolism. There are two distinct forms, both with episodic attacks of ataxia responsive to acetazolamide, with features reminiscent of periodic paralysis and suggestive of underlying ion channel abnormalities.

Episodic ataxia type 1 (EA1), an autosomal dominant disorder involving both the central and the peripheral nervous system, is characterized by attacks of ataxia and persistent myokymia. Episodes of ataxia, with gait imbalance and slurring of speech, occur spontaneously or can be precipitated by sudden movement, excitement or exercise. The attacks generally last from seconds to several minutes at a time and may recur many times a day. Myokymia, or muscle rippling resulting from motor nerve hyperexcitability, may be observed particularly around the eyes or in the small hand muscles. Subclinical rhythmic muscle activity may be demonstrated by electromyography. Acetazolamide, a carbonic anhydrase inhibitor, may be effective in preventing attacks of ataxia (Browne et al 1994). The age of onset, frequency of attacks, and severity may vary widely among family members.

Episodic ataxia type 2 (EA2) is an autosomal dominant disorder with episodes of markedly impaired truncal ataxia lasting hours to days with interictal eye movement abnormalities. Exertion and stress commonly precipitate the episodes. Often, the episodes of ataxia respond to acetazolamide (Ophoff et al 1996). In some individuals, there may be a gradual baseline ataxia with evidence of cerebellar atrophy. Affected patients may also have migraine; some even complain of basilar migraine (Ophoff et al 1996).

Genetics

Linkage analysis of several large pedigrees with EA1 mapped the disease locus to 12p13, near a cluster of three K^+ channel genes: *KCNA1*, *KCNA5* and *KCNA6* (Liu & Sommer 1994). Based on the clinical phenotype and its analogy to episodic disorders of muscle, ion channel genes were considered good candidate genes for EA1. Browne and colleagues (Browne et al 1994) serendipitously focused on *KCNA1*: the individual who led these investigators to exclude *KCNA5* and *KCNA6* later turned out to represent a phenocopy. *KCNA1*, which encodes Kv1.1, a delayed rectifier K^+ channel, has no intervening introns in its 1488bp sequence. Analysis of the single exon in *KCNA1* identified point mutations in four unrelated EA1 pedigrees (Browne et al 1994). These mutations fulfilled several criteria for disease-causing mutations: they segregated with the phenotype, involve highly conserved amino acid residues in the K^+ channel gene product, and were not found in controls. Additional mutations in the same gene were subsequently identified in other EA1 pedigrees (reviewed in Jen & Ptacek 2000). There is not an animal model for EA1 with missense mutations in *KCNA1*. Smart and colleagues (Smart et al 1998) generated Kv1.1 null mice that suffer from frequent spontaneous seizures but display no evidence of ataxia.

The disease locus in EA2 in several pedigrees was localized to chromosome 19p. Earlier, the disease locus of half of the pedigrees with familial hemiplegic migraine (FHM) was mapped to the same locus (Joutel et al 1993). FHM is a dominantly inherited subtype of migraine with aura, characterized by recurrent attacks of migrainous headache with ictal hemiparesis. Recovery is generally complete, with a normal intenctal physical examination. The onset is often early in life. Overlap in symptoms between hemiplegic migraine and basilar migraine suggests that hemiplegic migraine may be a form of basilar migraine.

A Ca^{2+} channel subunit gene mapping to 19p was an ideal candidate gene for EA2 and FHM. Ophoff and colleagues (Ophoff et al 1996) first defined the complex gene structure of *CACNA1A*, which spans 300 000 bp and consists of 47 exons that encode the α1A subunit with 2261 amino acids. These investigators then analysed the exons and flanking introns of *CACNA1A* and identified missense mutations in pedigrees affected with FHM and point mutations that result in premature stop or interfere with splicing in EA2 families.

Subsequently, small expansions of a polymorphic CAG repeat in the same gene were identified in some ataxic patients with a condition since designated SCA6, a dominantly inherited progressive cerebellar ataxia clinically indistinguishable from other dominant ataxic syndromes (Zhuchenko et al 1997). Some patients experience fluctuating symptoms, similar to episodic ataxia.

Yue et al (1997) identified a missense mutation in a family with severe progressive ataxia and superimposed episodes of vertigo and ataxia. Providing additional evidence supporting *CACNA1A* as the disease-causing gene in episodic ataxia, these researchers identified in a patient with episodic ataxia but no family history a *de novo* mutation that predicts a premature stop code.

Of note, mutations in genes encoding Ca^{2+} channel subunits have been identified in a number of recessive mouse mutants with ataxia. Homozygous point mutation in the α1A gene (P1802L, domain II P-region) causes epilepsy and ataxia in the mutant mouse tottering (*tg*) (reviewed in Jen & Ptacek 2000). Novel sequences in the intracellular C-terminus of the α1A subunit also resulted in a mutant mouse phenotype, leaner (tg^{la}), with ataxia and epilepsy (reviewed in Jen & Ptacek 2000). Burgess and colleagues reported that mutation of a Ca^{2+} channel gene with a predicted deletion of the highly conserved α1-binding motif in the $\beta4$ subunit was associated with seizures and ataxia in a mutant mouse, lethargic (*lh*) (reviewed in Jen & Ptacek 2000). Characterization of the genetic defects in the stargazer and waggler mutants lead to the identification of a neuronal Ca^{2+} channel γ subunit (reviewed in Jen & Ptacek 2000).

Congenital myasthenic syndromes

Congenital myasthenic syndromes are a group of rare, hereditary, non-immune-mediated disorders of neuromuscular transmission. Depolarization of the motor neuron activates presynaptic voltage-gated Ca^{2+} channels to allow Ca^{2+} ion influx to trigger the release of vesicles containing acetylcholine (ACh). ACh released into the synapse binds to nicotinic ACh receptors (nAChRs) clustered in endplate (EP) on the postsynaptic muscle surface membrane. The entry of cations through the activated nAChRs leads to depolarization (endplate potential), thus activating voltage-gated Na^+ channels and ultimately muscle contraction. ACh is cleared from the synaptic cleft by acetylcholinesterase (AChE).

Defects involving the presynaptic, synaptic, and postsynaptic components of the neuromuscular junction have been identified in different pedigrees affected with myasthenic symptoms. Postsynaptic dysfunction accounts for the majority of congenital myasthenic syndromes. A wealth of detailed studies of the physiologic, pharmacologic and molecular properties of nAChR greatly facilitated the clinical characterization and subsequent identification of nAChR dysfunction causing congenital myasthenia. In particular, in two pedigrees with congenital myasthenic syndrome, Engel et al (1996) performed electrophysiologic studies to demonstrate markedly prolonged EP potential, which led to the hypothesis that mutations in one or more nAChR channel subunit could underlie the abnormal kinetics observed. This hypothesis was subsequently confirmed by the identification of specific mutations in genes encoding different subunits of the nAChR in slow-channel congenital myasthenic syndrome (SCCMS). In addition, other mutations in nAChR subunit genes were associated with fast channel syndrome and nAChR deficiency.

nAChR

The nAChR complexes are ligand-gated channels expressed in skeletal muscles and brain. Five different subunits exist: $\alpha1$–9, $\beta1$–4, γ, δ and ε. These subunits are encoded by different genes on different chromosomes. They are homologous, each with a large N-terminal extracellular domain and four transmembrane regions, with a large intracellular loop between M3 and M4 (Fig. 3).

The subunit composition at the neuromuscular junction is fixed: $(\alpha1)_2(\beta1)\delta\gamma$ in the fetal form and $(\alpha1)_2(\beta1)\delta\varepsilon$ in the adult form. M2 from each subunit lines the channel pore. A leucine residue at homologous locations in each M2 segment forms a hydrophobic ring critical for channel gating. There are two binding sites for ACh, located at the interface between N-terminal hydrophilic domains of α-δ and α-ε or α-γ. The snake neurotoxin α-bungarotoxin irreversibly blocks these binding sites and has been useful in labelling and counting nAChR.

Binding of two ACh molecules induces conformational changes of the channel complex to an open state to allow mostly Na^+ but also Ca^{2+} to enter the cell. The channel may undergo further voltage-dependent conformational changes before it closes and ACh dissociates from the receptor channel complex.

Clinical features

Slow-channel congenital myasthenic syndrome (SCCMS) is an autosomal dominant hereditary condition characterized clinically by weakness, fatigability, and progressive muscle atrophy (Engel et al 1996). Patients generally present early, with poor head control and weak suck during infancy. Weakness of skeletal muscles, particularly bulbar, cervical and hand extensor muscles, may be present since birth. The age of onset and severity may vary among family members. Laboratory testing reveals no presence of anti-AChR antibodies. Morphological studies at the light microscopic level show focal endplate myopathy. Ultrastructural studies reveal degenerative changes in the basement membrane, postsynaptic membrane and the synaptic organelles of the neuromuscular junction. Electrophysiological testing demonstrates many abnormalities. At a normal neuromuscular junction, a single stimulus should elicit a single compound muscle action potential (CMAP). In contrast, a single stimulus in SCCMS may elicit multiple potentials with decremental response, in addition to the initial CMAP. On repetitive stimulation of motor nerves, there is a decremental electromyographic response. Intracellular recordings of biopsied muscles show prolonged decay of endplate potentials. Single channel recordings demonstrate increased channel opening compared to control.

Deficiency of nAChR is an autosomal recessive disorder causing congenital myasthenic syndrome with generalized weakness of early onset. The most striking abnormality in laboratory studies is the marked reduction in nAChR in the motor endplate measured by α-bungarotoxin labelling. The patients usually respond favourably to AChE inhibitors. Immunocytochemical studies using labelled antibodies directed against the γ subunit have successfully demonstrated the presence of the immature subunit at the neuromuscular junction. There may be an increased number of small endplates distributed over an increased span of muscle fibre surface. Electrophysiologic studies showed evidence of AChR with small conductance and slow kinetics characteristic of a γ-containing rather than ε-containing channel complex.

The low-affinity fast channel syndrome is a rare, recessive condition where patients generally present with moderately severe myasthenic symptoms from birth (Uchitel et al 1993, Ohno et al 1996). Electrophysiologic studies revealed decremental CMAP and very small MEPPs. Single channel recordings from biopsied muscle endplate showed infrequent, abnormally brief openings in response to ACh. Morphologic studies demonstrated normal end plate structure, with normal number of AChR and no evidence of myopathy.

Genetics

The abnormal 'slow channel' kinetics suggests abnormal channels, which could result from mutations in one or more subunit of the nAChR. Patients with electrophysiologically confirmed slow-channel phenotype were screened for mutations in genes encoding the $\alpha1$, $\beta1$, δ and ε subunits. Single strand conformation polymorphism analysis of PCR-amplified exons and flanking introns of all subunits of nAChR was performed. Direct sequencing of aberrant bands successfully identified mutations in different subunit genes.

Several groups have reported recessive, heteroallelic mutations in the ε subunit gene in patients with nAChR deficiency or the low-affinity fast channel syndrome.

Hereditary hyperekplexia: glycine receptor channel

Clinical manifestations

Hereditary hyperekplexia, or startle disease, is a rare, highly penetrant autosomal dominant disorder characterized by an exaggerated startle response and hypertonia. The normal startle reflex is a primitive reflex with a complex, stereotyped pattern of motor behaviour in response to unexpected sensory stimuli. The motor response generally consists of blinking, grimacing, flexion of neck and arms, and delayed abduction of the hand muscles. Patients with startle syndromes due to various neuropathologic conditions may present with abnormally exquisite sensitivity or abnormally violent motor response to sudden stimuli. In particular, studying hereditary hyperekplexia has helped elucidate the pathophysiologic and genetic basis of abnormal startle response.

A small kindred with 'emotionally precipitated drop seizures' described by Kirstein and Silvferskiold in 1958 was perhaps the first case report of hereditary hyperekplexia (Kirstein & Silfverskiold 1958). Suhren et al (1966) subsequently described a major form and a minor form of excessive response to sudden stimuli in a large Dutch pedigree spanning five generations. The symptoms that characterize the major form consisted of transient hypertonia during infancy, an exaggerated startle response with generalized stiffening causing falls but no loss of consciousness, repetitive limb jerking particularly at night, hyperekplexia and hesitant gait. Excessive startle without any associated symptoms characterizes the minor form.

Symptoms of the major form of hereditary hyperekplexia could present early, as in two unrelated patients who had unusual fetal movements during gestation, with sudden forceful jerking lasting from seconds to minutes that increased in severity in response to external stimuli (Leventer et al 1995). Shortly after birth both exhibited rigidity, nocturnal limb jerking and an exaggerated startle response, which are typical findings in the neonatal period in patients with hereditary hyperekplexia. Unexpected stimuli, such as noises or normal handling, could precipitate massive, generalized spasms of skeletal muscles to cause apnoea, cyanosis and even death, during infancy (Kurczynski 1983). Often mistaken for spastic quadriplegia, neonatal stiffness improves in early childhood (Andermann et al 1980). Delay in motor development is common because any sudden sound, touch, or movement could cause patients to stiffen and fall. Diffuse stiffness renders the patients completely powerless in breaking their fall. There is no loss of consciousness. Patients may develop an awkward slow wide-based gait because of fear of falling. Inguinal and abdominal hernia as well as generalized seizures independent of startle in patients with hyperekplexia have also been reported (Suhren et al 1966, Kurczynski 1983). Anxiety, stress, fear, sleep deprivation and menstruation exacerbate the symptoms.

Physical examination usually reveals diffusely increased muscle tone during infancy that normalizes with age. Tapping of the forehead results in excessive head retraction. Deep tendon reflexes are generally hyperactive. The preservation of consciousness and the absence of epileptiform discharges on EEG distinguish hyperekplexia from startle epilepsy (Andermann et al 1980).

Analysis of the electromyographic studies as well as evoked responses demonstrated the same startle pattern in control subjects as in patients with hyperekplexia, regardless of aetiology (Brown et al 1991, Matsumoto et al 1992). The acoustic startle pathway is not completely understood but is thought to involve mainly the caudal brainstem. Lesion and stimulation studies identified a circuitry, with signals travelling from the auditory nerve to the ventral cochlear nucleus, nuclei of the lateral lemniscus, nucleus reticularis pontis caudalis, reticulospinal tract, spinal interneurons and finally, the lower motor neurons to innervate skeletal muscles (Davis et al 1982). Lesions involving the startle circuit spanning the caudal brainstem and the spinal cord could lead to decreased startle. Excessive startle would indicate increased excitatory or decreased inhibitory input to the startle circuit.

Genetics

Linkage studies in several unrelated kindreds with autosomal dominant inheritance of hereditary hyperekplexia localized the disease gene to the distal portion of the long arm of chromosome 5 (Ryan et al 1992). This region

TABLE 1 Channelopathies

Disease	*Gene*	*Ion channel*
Hyperkalaemic periodic paralysis	*SCN4A*	Na^+ channel
Paramyotonia congenita	*SCN4A*	Na^+ channel
Hypokalaemic periodic paralysis 1	*CACNLA3*	Ca^{2+} channel
Hypokalaemic periodic paralysis 2	*SCN4A*	Na^+ channel
Myotonia congenita	*CLCN1*	Cl^- channel
Episodic ataxia with myokymia (type 1)	*KCNA1*	K^+ channel
Episodic ataxia with nystagmus (type 2)	*CACNA1A*	Ca^{2+} channel
Familial hemiplegic migraine	*CACNA1A*	Ca^{2+} channel
Spinocerebellar ataxia type 6	*CACNA1A*	Ca^{2+} channel
Congenital myasthenic syndrome	*CHRNA,B,E*	Acetylcholine receptor
Hereditary hyperekplexia	*GLRA1*	Glycine receptor
Long QT syndrome 1	*KCNQ1*	K^+ channel
Long QT syndrome 2	*HERG*	K^+ channel
Long QT syndrome 3	*SCN5A*	Na^+ channel
Long QT syndrome 4	*?*	?
Long QT syndrome 5	*KCNE1*	K^+ channel
Benign familial neonatal convulsions 1	*KCNQ2*	K^+ channel
Benign familial neonatal convulsions 2	*KCNQ3*	K^+ channel

contained genes encoding subunits of GABA receptor, glycine receptor, adrenergic receptor and glutamate receptor. Radiation hybrid mapping analysis demonstrated that only the gene encoding the $\alpha 1$ subunit of the inhibitory glycine receptor (GLRA1) was within the disease gene region (Shiang et al 1993). GLRA1 was a good candidate gene for hyperekplexia because glycinergic transmission mediates recurrent inhibition of pontomedullary reticular neurons as well as spinal motor neurons. Screening by denaturing gradient gel electrophoresis combined with direct sequencing uncovered two different missense mutations in exon 6 of GLRA1 in affected members from four families, substituting an uncharged amino acid (leucine or glutamine) for arginine, a charged residue. Additional mutations in GLRA1 were subsequently identified in other hereditary and sporadic cases of hyperekplexia. (Table 1)

Mice are excellent animal models for studying hereditary hyperekplexia in humans. When given sublethal doses of strychnine, which is a competitive glycine receptor antagonist, normal mice display hypertonia and an exaggerated startle reflex, reminiscent of the cardinal features of hyperekplexia in humans.

Furthermore, two autosomal recessive mouse mutants, spastic (*spa*) and spasmodic (*spd*), exhibit identical phenotypes with striking similarities to STHE in human. Phenotypically normal at birth, the affected animals develop rigidity and fall in response to sudden tactile or acoustic stimuli after the second postnatal week. A lethal mutant, oscillator (*ot*), displays more severe symptoms and may be allelic to *spd*.

The phenotypic similarities between *spd* and hereditary hyperekplexia suggest that mutations in glycine receptor gene may be responsible for the *spd* phenotype. The disease locus of *spd* mapped to a small region on mouse chromosome 11 that was homologous to the hereditary hyperekplexia-containing region on human chromosome 5q (Lane et al 1987, Buckwalter et al 1993). Direct sequencing showed that *spd* indeed was the mouse homologue of hyperekplexia and a specific missense mutation (G534T) was identified in *spd*, which would result in an Ala52Ser substitution in the mouse glycine receptor $\alpha 1$ subunit, sharing 99% peptide sequence homology with the adult human orthologue (Ryan et al 1994). Notably, the expression of glycine receptor isoforms is developmentally regulated, switching from the neonatal to the adult isoform around the second postnatal week in mice. The neonatal isoform of the glycine receptor contains $\alpha 2$ subunits only, while the adult isoform appears to be composed of three $\alpha 1$ subunits and two β subunits. That *spd* mice are phenotypically normal until after the second postnatal week is consistent with the expression of the mutation-containing $\alpha 1$ subunit gene of the adult isoform of glycine receptor.

Other paroxysmal neurological disorders

A growing list of paroxysmal neurological disorders has been shown to result from mutations in ion channels. The identification of specific mutations has made diagnosis and patient classification possible. Physiologic characterization of mutant channels will continue to reveal new functional domains. The finding that mutations in the auxiliary subunits may have the same clinical manifestations as mutations in the main, pore-forming channel subunits indicates that defects in any component (upstream or downstream) with which the channel protein interacts can lead to dysfunction. That mutations in distinct proteins may produce similar phenotypes provides insight to the functional role of the different channel proteins that must be carefully coordinated to carry out proper neurological activities. In the case of congenital myasthenic syndromes, defects involving the presynaptic, synaptic, or postsynaptic component of the neuromuscular junction may interfere with normal neuromuscular transmission. It is intriguing that different mutations in the same gene produce different phenotypes (although with some overlap), as illustrated by the allelic disorders involving mutations in *SCN4A* and *CACNA1A*. Phenotypic variability among individuals within the same pedigree or different pedigrees with the same mutation suggests that other genetic and/or environmental factors must contribute to the phenotypic expression.

Understanding the phenotypic expression of the rare, monogenic channelopathies presented in this chapter may help elucidate similar mechanisms in other paroxysmal neurological disorders, such as familial paroxysmal dyskinesia, migraine and epilepsy. Familial paroxysmal dyskinesia is a rare group of disorders characterized by episodic involuntary hyperkinetic movement (Demirkiran & Jankovic 1995). A familial paroxysmal choreoathetosis syndrome associated with progressive spasticity mapped to chromosome 1p (Auburger et al 1996), while paroxysmal dystonic choreoathetosis not associated with spasticity mapped to chromosome 2q34 (Fouad et al 1996). Ion channels are candidate genes for these episodic movement disorders.

Since migraine shares many features with known channelopathies, ion channels are likely sites of genetic defects in this heterogeneous and possibly polygenic disorder. In addition to *CACNA1A* mutations discussed in this chapter, two other loci on chromosome 1q have been identified in other pedigrees with familial hemiplegic migraine, which is a rare form of migraine with aura (Ducros et al 1997, Gardner et al 1997).

Epilepsy also shares many features with known channelopathies; indeed, mutations have been identified in both ligand-gated and voltage-gated ion channel genes in different pedigrees with epilepsy (Steinlein et al 1995, 1997, Biervert et al 1998, Charlier et al 1998, Singh et al 1998).

With the Human Genome Project and the development of DNA chip technology well under way, soon we will be able to easily identify genes and mutations. One of the challenges that we face is to understand how these ion channel mutations are translated into paroxysmal neurological disturbances as well as fixed neurological deficits. Studying the biophysical properties of individual ion channels is a good start. Other processes such as gene expression, alternative splicing, assembly, subcellular localization, modulation by protein kinases/ phosphatases, interaction with synaptic machinery and cytoskeletal elements, and excitotoxicity are incompletely understood but clearly important in elucidating the underlying disease-causing mechanism and developing rational treatment.

Acknowledgements

This work was funded by NIH and the Howard Hughes Medical Institute.

References

Abdalla JA, Casley WL, Cousin HK et al 1992 Linkage of Thomsen disease to the T-cell-receptor beta (TCRβ) locus on chromosome 7q35. Am J Hum Genet 51:579–84

Andermann FD, Keene L, Andermann E, Quesney LF 1980 Startle disease or hyperekplexia: further delineation of the syndrome. Brain 103:985–997

Auburger G, Ratzlaff T, Lunkes A et al 1996 A gene for autosomal dominant paroxysmal choreoathetosis/spasticity (CSE) maps to the vicinity of a potassium channel gene cluster on chromosome 1p, probably within 2 cM between D1S443 and D1S197. Genomics 31:90–94

Becker PE 1971 Genetic approaches to the nosology of muscular disease: myotonias and similar diseases. Birth Defects Orig Artic Ser 7:52–62

Biervert C, Schroeder BC, Kubisch C et al 1998 A potassium channel mutation in neonatal human epilepsy. Science 279:403–406

Brown P, Rothwell JC, Thompson PD, Britton TC, Day BL, Marsden CD 1991 The hyperekplexias and their relationship to the normal startle reflex. Brain 114:1903–1928

Browne DL, Gancher ST, Nutt JG et al 1994 Episodic ataxia/myokymia syndrome is associated with point mutations in the human potassium channel gene, KCNA1. Nat Genet 8:136–140

Buckwalter MS, Testa CM, Noebels JL, Camper SA 1993 Genetic mapping and evaluation of candidate genes for spasmodic, a neurological mouse mutation with abnormal startle response. Genomics 17:279–286

Charlier C, Singh NA, Ryan SG et al 1998 A pore mutation in a novel KQT-like potassium channel gene in an idiopathic epilepsy family. Nat Genet 18:53–55

Davis M, Gendehnan DS, Tischler MD, Gendelman PM 1982 A primary acoustic startle circuit: lesion and stimulation studies. J Neurosci 2:791–805

Demirkiran M, Jankovic J 1995 Paroxysmal dyskinesias: clinical features and classification. Ann Neurol 38:571–579

Ducros A, Joutel A, Vanedi K et al 1997 Mapping of a second locus for familial hemiplegic migraine to 1q21-q23 and evidence of further heterogeneity. Ann Neurol 42:885–890

Engel AG, Ohno K, Milone M et al 1996 New mutations in acetylcholine receptor subunit genes reveal heterogeneity in the slow-channel congenital myasthenic syndrome. Hum Mol Genet 5:1217–1227

Fontaine B, Khurana TS, Hoffman EP et al 1990 Hyperkalemic periodic paralysis and the adult muscle sodium channel alpha-subunit gene. Science 250:1000–1002

Fontaine B, Vale-Santos J, Jurkat-Rott K et al 1994 Mapping of the hypokalaemic periodic paralysis (HypoPP) locus to chromosome 1q31-32 in three European families. Nat Genet 6:267–272

Fouad GT, Servidei S, Durcan S, Bertini E, Ptacek LJ 1996 A gene for familial paroxysmal dyskinesia (FPD1) maps to chromosome 2q. Am J Hum Genet 59:135–139

Gardner K, Barmada MM, Ptacek LJ, Hoffman EP 1997 A new locus for hemiplegic migraine maps to chromosome 1q31. Neurology 49:1231–1238

George AL Jr, Sloan Brown K, Fenichel GM, Mitchell GA, Spiegel R, Pascuzzi RM 1994 Nonsense and missense mutations of the muscle chloride channel gene in patients with myotonia congenita. Hum Mol Genet 3:2071–2072

Jackson CE, Barohn RJ, Ptacek LJ 1994 Paramyotonia congenita: abnormal short exercise test, and improvement after mexiletine therapy. Muscle Nerve 17:763–768

Jen J, Ptacek LJ 2000 Channelopathies: episodic disorders of the nervous system. In: Schriver CR, Beaudet AL, Sly WS, Valle D, Vogelstein B, Childs B (eds) Metabolic and molecular bases of inherited disease. McGraw-Hill, New York, p 5223–5238

Joutel A, Bousser MG, Biousse V et al 1993 A gene for familial hemiplegic migraine maps to chromosome 19. Nat Genet 5:40–45

Kirstein L, Silfverskiold B 1958 A family with emotionally precipitated 'drop seizures'. Acta Psychiatr Neurol Scand 33:471–475

Koch MC, Steinmeyer K, Lorenz C et al 1992 The skeletal muscle chloride channel in dominant and recessive human myotonia. Science 257:797–800

Kurczynski TW 1983 Hyperekplexia. Arch Neurol 40:246–248

Lane PW, Ganser AL, Kerner AL, White WF 1987 Spasmodic, a mutation on chromosome 11 in the mouse. J Hered 78:353–356
Leventer RJ, Hopkins IJ, Shield LK 1995 Hyperekplexia as cause of abnormal intrauterine movements. Lancet 345:461
Liu Q, Sommer SS 1994 Parameters affecting the sensitivities of dideoxy fingerprinting and SSCP. PCR Methods Appl 4:97–108
Matsumoto J, Fuhr P, Nigro M, Hallett M 1992 Physiological abnormalities in hereditary hyperekplexia. Ann Neurol 32:41–50
McClatchey AI, Van den Bergh P, Pericak-Vance MA et al 1992 Temperature-sensitive mutations in the III-IV cytoplasmic loop region of the skeletal muscle sodium channel gene in paramyotonia congenita. Cell 68:769–774
Ohno K, Wang HL, Milone M et al 1996 Congenital myasthenic syndrome caused by decreased agonist binding affinity due to a mutation in the acetylcholine receptor epsilon subunit. Neuron 17:157–170
Ophoff RA, Terwindt GM, Vergouwe MN et al 1996 Familial hemiplegic migraine and episodic ataxia type-2 are caused by mutations in the Ca^{2+} channel gene CACNL1A4. Cell 87:543–552
Ptacek LJ 1998 The familial periodic paralyses and non-dystrophic myotonias. Am J Med 105:58–70
Ptacek LJ, George AL Jr, Griggs RC et al 1991a Identification of a mutation in the gene causing hyperkalemic periodic paralysis. Cell 67:1021–1027
Ptacek LJ, Tyler F, Trimmer JS, Agnew WS, Leppert M et al 1991b Analysis in a large hyperkalemic periodic paralysis pedigree supports tight linkage to a sodium channel locus. Am J Hum Genet 49:378–382
Ptacek LJ, Trimmer JS, Agnew WS, Roberts JW, Petajan JH, Leppert M 1991c Paramyotonia congenita and hyperkalemic periodic paralysis map to the same sodium-channel gene locus. Am J Hum Genet 49:851–854
Ptacek LJ, Tawil R Griggs RC, Storvick D, Leppert M 1992a Linkage of atypical myotonia congenita to a sodium channel locus. Neurology 42:431–433
Ptacek LJ, George AL Jr, Barchi RL et al 1992b Mutations in an S4 segment of the adult skeletal muscle sodium channel cause paramyotonia congenita. Neuron 8:891-897
Ptacek LJ, Tawil R, Griggs RC et al 1994a Sodium channel mutations in acetazolamide-responsive myotonia congenita, paramyotonia congenita, and hyperkalemic periodic paralysis. Neurology 44:1500–1503
Ptacek LJ, Tawil R, Griggs RC et al 1994b Dihydropyridine receptor mutations cause hypokalemic periodic paralysis. Cell 77:863–868
Rojas CV, Wang JZ, Schwartz LS, Hoffman EP, Powell BR, Brown RH Jr 1991 A Met-to-Val mutation in the skeletal muscle Na+ channel alpha-subunit in hyperkalaemic periodic paralysis. Nature 354:387–389
Ryan SG, Sherman SL, Terry JC, Sparkes RS, Torres MC, Mackey RW 1992 Startle disease, or hyperekplexia: response to clonazepam and assignment of the gene (*STHE*) to chromosome 5q by linkage analysis. Ann Neurol 31:663–668
Ryan SG, Buckwalter MS, Lynch JW et al 1994 A missense mutation in the gene encoding the alpha 1 subunit of the inhibitory glycine receptor in the spasmodic mouse. Nat Genet 7:131–135
Shiang R, Ryan SG, Zhu YZ, Hahn AF, O'Connell P, Wasmuth JJ 1993 Mutations in the alpha 1 subunit of the inhibitory glycine receptor cause the dominant neurologic disorder, hyperekplexia. Nat Genet 5:351–358
Singh NA, Charlier C, Stauffer D et al 1998 A novel potassium channel gene, *KCNQ2*, is mutated in an inherited epilepsy of newborns. Nat Genet 18:25–29
Smart SL, Lopantsev V, Zhang CL et al 1998 Deletion of the K(V)1.1 potassium channel causes epilepsy in mice. Neuron 20:809–819

Steinlein OK, Mulley JC, Propping P et al 1995 A missense mutation in the neuronal nicotinic acetylcholine receptor alpha 4 subunit is associated with autosomal dominant nocturnal frontal lobe epilepsy. Nat Genet 11:201–203

Suhren O, Brruyn G et al 1966 Hyperekplexia: a hereditary startle syndrome. J Neurol Sci 3:577–605

Trudell RG, Kaiser KK Griggs RC 1987 Acetazolamide-responsive myotonia congenita. Neurology 37:488–491

Tyler FH, Stephens FE 1951 Studies in disorders of muscle VII. Clinical manifestations and inheritance of a type of periodic paralysis without hypopotassimia. J Clin Invest 30:492–502

Uchitel O, Engel AG, Walls TJ, Nagel A, Atassi MZ, Bril V 1993 Congenital myasthenic syndromes: II. Syndrome attributed to abnormal interaction of acetylcholine with its receptor. Muscle Nerve 16:1293–1301

Yue Q, Jen JC, Nelson SF, Balo RW 1997 Progressive ataxia due to a missense mutation in a calcium channel gene. Am J Hum Genet 61:1078–1087

Zhang J, George AL Jr, Griggs RC et al 1996 Mutations in the human skeletal muscle chloride channel gene (*CLCN1*) associated with dominant and recessive myotonia congenita. Neurology 47:993–998

Zhuchenko O, Bailey J, Bonnen P et al 1997 Autosomal dominant cerebellar ataxia (SCA6) associated with small polyglutamine expansions in the α1A-voltage-dependent calcium channel. Nat Genet 15:62–69

DISCUSSION

Horn: Some of the environmental factors that induce periodic paralysis, such as changes in K^+ concentration or temperature, don't seem to work in *invitro* studies. Have they worked at all, and do you have any idea as to how K^+ or temperature is affecting the function of the Na^+ channel?

Ptacek: No. I think that the German group did show a K^+ effect when they studied patient muscle *in vitro*. Clearly, it is more complex than a direct effect of K^+ or temperature on channels in a transfected human kidney cell, for example. It is a fascinating area that we hope one day to understand.

Raman: In those studies, is there any description of what the currents through the channels look like in the native tissue, or how they are altered with the various mutations?

Ptacek: Steve Cannon has studied cultured myoblasts from a patient with the M1592V mutation. He showed that while normal channels gave him two different modes (a fast and slow mode) the myoblasts in culture also gave him two modes but with a higher percentage of the channels gating in the slow mode. He hypothesized that the small number of channels that were gating in the slower mode contributed to a persistent Na^+ conductance which fed back because of depolarizing effects on the membrane to activate both wild-type and deactivated mutant Na^+ channels, thus leading to a spiralling event of more and more depolarization.

Horn: Did he look at the effect of K^+ on those cells?

Cummins: He did. In the human tissue there seemed to be a striking K^+ sensitivity. However, the same mutation did not seem sensitive to K^+ when expressed in HEK293 cells. But in the myoblasts you have both the wild-type and the mutated allele, plus probably some cardiac Na^+ channels starting to pop up. One difficulty is that obtaining diseased human tissue is rare. It is going to be even harder to obtain diseased tissue for electrophysiological studies with the neuronal disorders. You can't easily study them in an isolated expression system if there are so many factors needed to precipitate the abnormal activity.

Ptacek: One interesting point here is that it is a lot easier to get muscle tissue than it is to get brain tissue. I think this is one of the reasons. We can get tissue from these patients, but it is hard work. It is also difficult to study native muscle. It would be interesting to biopsy patients with different mutations, to characterize functional consequences and to compare these with those seen in heterologous expression systems.

Raman: Are there mouse models that show a similar phenotype?

Cummins: People have tried to make them, but I've not seen one that is successful.

Keynes: Is this the same myotonia that goats get?

Ptacek: No, they have a Cl^- channel mutation. This is a separate story.

Segal: In the temperature-dependent febrile seizures, the seizures tend to occur typically during a rise in temperature. *In vitro* I have seen a similar effect in solitary neurons in microisland cultures where it is not the absolute temperature but rather the change in temperature that increases the excitability (M. Segal, unpublished results). Could this be the case for some of the muscle disorders?

Ptacek: I'm not aware that people have looked at this. Dick Horn's group has done some nice work looking at the paramyotonia mutations.

Horn: We changed the temperature on the same cell that was transfected with the paramyatonia congenita mutation, and we didn't see any unusual effects.

Ptacek: Presumably you can study these cells at physiological temperature but this is more difficult because the kinetics are so fast.

Horn: You can, but we didn't go up to 37 °C.

Ptacek: One hypothesis is that it might be something as simple as changes in membrane fluidity, which could then affect the way that the channel can make conformational shifts. Another hypothesis is that other temperature-sensitive processes such as enzymes that modulate the channels could be altered in subtle ways.

Noebels: Is anything known about whether external K^+ ions could alter the Na^+ pore in a way that transmembrane voltage wouldn't?

Bean: I think Steve Cannon's current hypothesis of the effect of K^+ has to do with the K^+ concentration in the T tubules of the muscle. The idea is that when there is a lot of activity, K^+ coming out through the K^+ channel builds up in the T tubules. A crucial factor in the behaviour of the overall muscle is the K^+ concentration in the T tubules, because this can then depolarize the membrane and bring the Na^+ channels towards threshold. I think it is an indirect effect: high K^+ biases that mechanism.

Noebels: Clinically, if you take a patient and actually bolus them with K^+, or insulin to remove extracellular K^+, how long does it take for them to develop symptoms? Is this something that shows as a chronic change in the muscle, and when the person is weak and the physician finds their K^+ is low? How long has it been low for? Could the muscle cells have changed pump activity or other interesting properties so that we are using a false correlation of a membrane state with ion concentration? How fast is the reaction with K^+?

Ptacek: It differs from patient to patient, but it is not immediate. It generally takes about 30–60 minutes. We can't give a huge bolus of K^+, since this can be dangerous for the patient.

Noebels: So it is relatively quick.

Crill: In the hyerkalaemic paralysis, since you can make muscle contract without any extracellular Ca^{2+}, isn't it generally thought that Ca^{2+} isn't flowing through the Ca^{2+} channel?

Ptacek: No, Ca^{2+} is definitely going through the channel, but it is not thought that this Ca^{2+} is important for electrochemical coupling. Rather, the channel is serving as a voltage sensor for the ryanodine receptor and slow release channel. It clearly is a Ca^{2+} channel, however, and presumably that Ca^{2+} flux through the channel is doing something, although it is not EC coupling.

Bean: One idea is that Ca^{2+} entry through the Ca^{2+} channel is necessary for long-term loading of the SR.

Crill: In the hyerkalaemic periodic paralysis, when they are paralysed are their muscles depolarized?

Ptacek: They are completely inexcitable and depolarized. It can't be simply that there are levels of Ca^{2+} large enough to depolarize the membrane because the patient would be in tetany. They are inexcitable, so it is not that you are disrupting EC coupling.

Crill: Has anyone looked at the acquired form of hypokalaemic periodic paralysis in orientals?

Ptacek: Again, those are pretty rare. The patients are thyrotoxic and they have paralysis just like the hypokalaemics, except that when you ablate their thyroid, the thyrotoxic periodic paralysis resolves. There was one small family reported. I believe that this is a hereditary disorder, but we don't see it as a hereditary disorder much because it is a predisposing mutation and you also have to be thyrotoxic for the paralysis to manifest. We've looked for Ca^{2+} channel mutations in those patients but haven't found any.

Crill: The phenotype of the periodic paralysis is virtually identical.

Ptacek: The one thing that is different is that their K^+ levels tend to be much lower. The familial hypokalaemic patients would typically have a K^+ level of 2.1–3.2 during an attack. If I see a patient that has a K^+ level of 1.7–1.9 (which is alarmingly low), the first thing I think of is thyrotoxic periodic paralysis, because they tend to be really low. I don't know precisely why this is the case.

Isom: Rocky Kass identified a long QT mutation that interrupted the association of α with $\beta 1$, which you could imagine might have effects on gating mode or voltage dependence. Have you ever seen anything like this, or have you found any mutations in the β subunits?

Ptacek: We have looked hard. After reading your paper on the β subunit, I became interested because the β subunit is a great candidate for Andersen's syndrome: the same β subunit is used by cardiac and skeletal Na^+ channels. So we looked hard in the β subunit and couldn't find any mutations. I was sure that this was going to be the Andersen's gene, but it appears that it isn't. The only β subunit mutation that I am aware of is the epilepsy mutation identified by John Mulley's group.

Isom: $\beta 1$ is everywhere. You'd expect those people to have some defects in the heart or skeletal muscle, but they don't present that way. Have you ever seen that? Is there any overlap there?

Ptacek: No, I haven't. It is an interesting question. Without the β subunit, the α subunit physiology looks just like the mutations with the persistent Na^+ conductance.

Bean: What are their symptoms?

Isom: Febrile infantile seizure, which reappears later in life as an adult seizure that is afebrile.

Ptacek: This gets back to issues that were raised earlier: we are interested in the primary determinants and variations in those genes, but then this also has to do with the spatiotemporal expression of that gene and others that can compensate for it. It may be that other β subunits or perhaps even alternate isoforms of α subunits might interact in different ways such that we don't see a phenotype manifest in the brain but we do in the heart, or vice versa.

Isom: It could be the particular cell adhesion events in the brain, heart and skeletal muscle, such that adhesion is interrupted in the brain but not the other tissues.

Meisler: I share your view that the interaction of polymorphic variants in these Na^+ channels could have profound effects on what we consider as psychiatric function. It is going to be very interesting: one can imagine in the future that neurological exams will include diagnostic chips for scanning mutations in all the channels simultaneously. We have defined two common haplotypes in the human SCN1A channel. These common variants could be analysed in your population. I wanted to ask particularly about the Roman family. It looked like you had linkage.

Ptacek: That is linked to 2q. There was another beautiful candidate gene, which encodes a chloride/bicarbonate exchanger. There is a good physiological argument about how this gene could cause that disease. As so many great hypotheses do, it went down in flames.

Crill: What is the current idea about how hyperkalemic patients can prevent the depolarization by continued exercise?

Ptacek: During exercise these patients don't get an attack. One idea is that increased bloodflow is more able to deal with and buffer accumulating K^+ levels. If you then continue to exercise gently and cool down more slowly this might be able to deal with the increased K^+ without any rebound increase later on. It is not very satisfying, but it seems to be the best explanation at this time.

Sodium channel gene expression and epilepsy

Jeffrey L. Noebels

Developmental Neurogenetics Laboratory, Department of Neurology, Baylor College of Medicine, One Baylor Plaza, Houston, TX 77030, USA

Abstract. Na^+ channelopathies that prolong membrane depolarization lead to neuronal bursting, abnormal network synchronization, and various patterns of episodic neurological disorders, including epilepsy. Two distinct pathways exist for generating epileptic phenotypes based on inherited disorders of voltage-gated Na^+ ion channels. The first pathway is direct, involving mutations in genes encoding the pore-forming $\alpha 1$ and regulatory β subunits of the channel that directly alter current amplitude or kinetics. These mutations favour repetitive firing and network hyperexcitability, although often the circuits most vulnerable to functional alterations are not easy to identify and the emergent clinical phenotypes are difficult to predict. The second pathway involves mutation of other genes that lead to downstream modifications in Na^+ channel expression. Two clinically relevant examples of localization-related vulnerability in brain are described that illustrate how specific phenotypes arise from both direct and secondary pathways. *Selective expression* of the cardiac SCN5A channel within limbic regions of brain may explain why mutation of the gene for this tetrodotoxin-insensitive current may be associated with seizures. *Ectopic expression* of type II Na^+ channels along axonal internodes in hypomyelinated brain may reveal why deletion of the myelin basic protein gene leads to subcortical seizure patterns. Analysis of these models offers insight into developmental processes that control the cellular expression and plasticity of Na^+ channel genes, and will help to clarify mechanisms of hereditary Na^+ channel-based epileptogenesis.

2002 Sodium channels and neuronal hyperexcitability. Wiley, Chichester (Novartis Foundation Symposium 241) p 109–123

The voltage-gated Na^+ channel provided one of the earliest molecular targets of epilepsy therapy, however, the idea that inherited defects prolonging the channel open state might play a role in its pathogenesis was slower in coming. To be sure, all cellular models of epileptogenesis focus on the prolonged neuronal depolarization that leads to the paroxysmal depolarizing shift (PDS), the cellular hallmark of abnormal network synchronization in epileptic brain. The PDS features a giant (>200 mV), sustained (150–300 ms) envelope of depolarization giving rise to tetrodotoxin (TTX)-sensitive Na^+ spike bursting. However, the Na^+ spike-firing thresholds and kinetics of action potential electrogenesis recorded at the soma in these same neurons appear relatively unaffected in most models, and experimental convulsants in use to produce synchronous discharges have concentrated instead on pharmacological blockade of repolarizing K^+, Cl^- currents, or on ligand-gated neurotransmitter receptor channels. Even the use of veratridine, a toxin that prolongs Na^+ channel opening during sustained depolarization by abolishing inactivation, is a relatively recent tool to reproduce the PDS response in brain networks (Otoom et al 1998).

The discoveries that hereditary Na^+ channelopathies can be associated with episodic hyperexcitability in *Drosophila* and human muscle disease (Loughney et al 1989, Ptacek et al 1991), and the identification of at least 13 genes for the heteromeric Na^+ channel complex (Goldin 2001) ultimately set the stage for a formulation of how specific excitability phenotypes expressed within the CNS might be inherited as disorders of Na^+ ion channels. This realization prompted the evaluation of Na^+ ion channel subunit genes as candidate disease loci in human positional cloning studies of epilepsy pedigrees, with the result that two naturally occurring epilepsy syndromes have so far been associated with Na^+ channel mutations that prolong depolarization. Further associations of epilepsy phenotypes with mutations of other members in this gene family are certain to emerge.

The numerical diversity of Na^+ channel genes and their variable expression patterns increase the probability that the anticipated channel subunit mutations will generate multiple CNS excitability profiles linked to distinct clinical seizure types. Each of these selective excitability increases in specific axonal pathways may act to synchronize large neuronal populations by synaptic or non-synaptic mechanisms. This review considers the anatomical aspects of Na^+ channel involvement in epileptogenesis, namely, how the site of expression of the Na^+ channel gene, both within the neuron and within specific brain pathways, acts as an important phenotypic determinant of the disorder.

Inherited Na^+ channelopathies and epileptic phenotypes

Two examples of inherited human mutations of Na^+ channel genes associated with epilepsy, including both the pore-forming $\alpha 1$ subunit, as well as the regulatory transmembrane β subunits have now been identified (Wallace et al 1998, Escayg et al 2000), and are reviewed in this volume and elsewhere (Steinlein & Noebels 2000).

Generalized epilepsy with febrile seizures plus (GEFS+1 and 2), syndromes classified by the presence of multiple seizure types both in the affected individual and within the extended disease pedigree, arise from point mutations that change function of the $\beta 1$ and $\alpha 1$ subunits, respectively. Although similarly classified, the mutations alter excitability in two distinct ways. The mutant C121W $\beta1$ subunit in GEFS+1 shows a loss-of-function phenotype that prolongs membrane repolarization when coexpressed with RBII $\alpha1$ subunits in oocytes. $\beta1$ subunits interact with multiple $\alpha1$ subtypes, and show overlapping expression patterns in the CNS (Isom 2002, this volume). Interestingly, this molecular promiscuity provides the regulatory subunit mutation with the capacity to create a complex channelopathy that alters neuronal excitability by changing the kinetics of multiple $\alpha1$ subunit types expressed differentially throughout the brain; a mechanism which may provide one explanation for the appearance of multiple seizure types in affected individuals. Once mutant mouse models of this defect become available, it will be instructive to determine where on the cell, in which cellular pathways, and under what conditions the *SCNB1* repolarization defect effectively results in epileptogenesis.

The same explanation may not be able to account for the various seizure types in the GEFS+2 syndrome, which involves a simple channelopathy consisting of either a C2624T or a G4943A substitution affecting only the *SCN1A* channel $\alpha1$ subunit. While no direct electrophysiology of mutant neurons is yet available, both mutations lie within the S4 transmembrane segments of domains 2 and 4 of the channel, and similar mutations introduced into domain 4 of either the *SCN2A* or *SCN5A* channels decrease channel inactivation (Kühn & Greeff 1999) or are associated with LQT (Wattanasirichaigoon et al 1999) while in *SCN4A* channels, only subtle lengthening of the time course of inactivation and a small reduction in voltage dependence were seen, without evidence of persistent Na^+ currents (Alekov et al 2000). Thus the multiple seizure types that arise in GEFS+2 individuals result from a diffuse but molecularly uniform change in SCN1A channel behaviour throughout the cerebral cortex, suggesting that other genetic, developmental, or environmental factors are more likely to account for phenotypic variability in this syndrome.

In addition to these reports, a second candidate gene for Na^+ channel $\alpha1$ subunit involvement in epilepsy has arisen due to the localization of *SCN5A* in brain, and the coexistence of human mutations in this gene that prolong depolarization that are associated with episodic cardiac and seizure phenotypes.

Restricted localization of SCN5A to brain limbic pathways

In a search for genes underlying TTX-insensitive Na^+ currents (White et al 1993) in brain, Hartmann et al (1999) localized the pattern of *SCN5A* gene expression in mammalian brain. Northern analysis of total brain RNA had failed to demonstrate expression of *SCN5A* mRNA. However, the more sensitive PCR amplified low levels of *SCN5A* cDNA from rat cortex and neonatal spinal cord (Yarowsky et al

1991, Black et al 1998). *In situ* hybridization and RT-PCR analysis revealed selective expression of *SCN5A* mRNA in the adult limbic forebrain. Autoradiograms of antisense probes for *SCN5A* revealed a sharply demarcated expression pattern in deep layers of the piriform cortex and within subcortical limbic circuitry (Fig. 1). Maximal specific labelling was observed throughout lateral septal nuclei, the bed nucleus of the stria terminalis, the piriform cortex, amygdala and periamygdaloid cortex and medial hypothalamic nuclei. Very low levels of *SCN5A* transcript were detected in the hippocampal formation, comparable to levels seen in other brain regions.

The restricted pattern of SCN5A is novel and contrasts with the more diffuse expression of other Na^+ ion channel $\alpha1$ subunits (Felts et al 1997, Whitaker et al 2000), predicting a specific but as yet undetermined role in regulating excitability and synchronization within these circuits. This uneven spatial distribution pattern of Na^+ channel $\alpha1$ subunit expression, mirrors a similar range of expression patterns of $\alpha1$ subunits for voltage-gated Ca^{2+} channels.

Neurons expressing *SCNA5* comprise a synaptically linked pathway from the piriform cortex and bed nucleus of the stria terminalis to all subdivisions of the amygdaloid nuclear complex, and reciprocal connections from the septofimbrial and septohypothalmic nuclei to the hippocampus and hypothalamus. The lateral septum receives a major, topographically organized input from the hippocampus, and projects back to this region via cholinergic and GABAergic neurons of the medial septum. The lateral septum is also a major relay of hippocampal output to hypothalamic nuclei, and may integrate endocrine responses with specific classes of emotional behaviour. The pattern of expression suggests that *SCN5A* contributes to neuronal signalling linking diverse functions, including olfactory perception, attention and spatial memory processes, and autonomic responses. Of potential interest is that each of these circuits displays prominent intrinsic oscillations (Bland & Colom 1993, Pape et al 1998).

The precise contribution of *SCN5A* channels to normal oscillatory behaviour within these pathways will require further study. However, functional defects arising in human *SCN5A* mutant phenotypes suggest an important pathophysiological role for this current in epilepsy. Mutations in the human *SCNA5* gene produce the cardiac long QT syndrome LQT3, characterized by prolonged myocardial depolarization with episodic tachyarrythmias (Wang et al 1995) and idiopathic ventricular fibrillation (IVF) (Chen et al 1998). Biophysical analysis of LQT3 mutations in heterologous systems reveal gating defects leading to sustained, non-inactivating currents. In the IVF phenotype, two classes of mutations in the coding region of *SCN5A* have been identified, including a missense mutation that recovers from inactivation more rapidly than normal, and splice-donor or frameshift mutations that cause loss of Na^+ current. Other LQT mutations may interfere with the $\beta1$ subunit interaction site (An et al 1998). The net

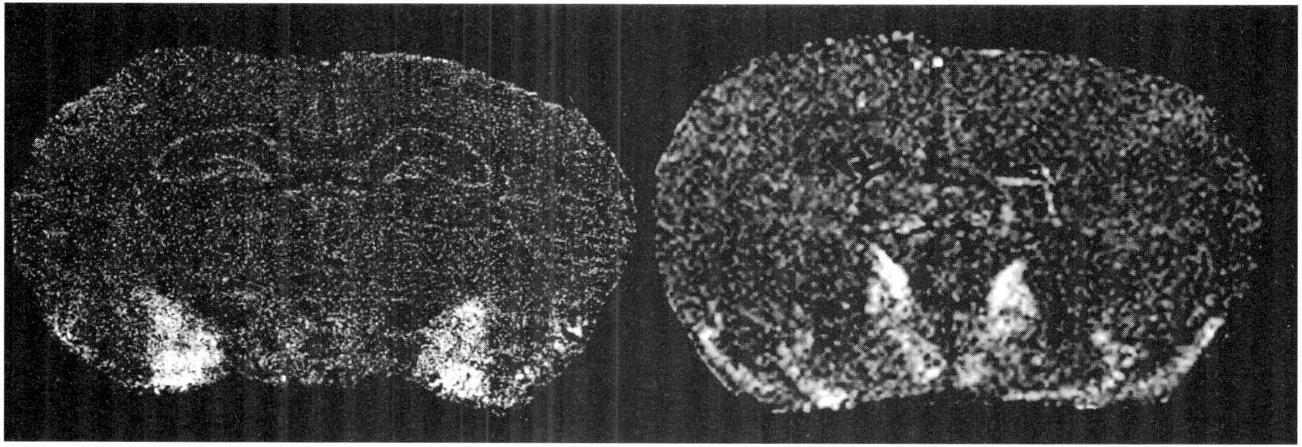

FIG. 1. Autoradiograms of labelled antisense probe hybridized to coronal sections of adult rat brain reveal selective localization of *SCNA5A* mRNA in limbic regions of rat brain. Strong expression of *SCN5A* is shown in amygdalar nuclei (left) and piriform cortex (right). (From Hartmann et al 1999.)

result of these defects is to disrupt pacemaking properties in ways that favour sudden fibrillation of the cardiac syncytium.

SCN5A mutations are likely to contribute to similar instabilities of neuronal rhythmic firing within limbic networks. The presence of verified epileptic seizures in individuals with LQT mutations (Herman et al 1992, Pacia et al 1994) supports the possibility that a subset of inherited epilepsy may arise from *SCN5A* channelopathies. Piriform cortex is the most susceptible of all forebrain regions to the induction of seizures, and regulates excitability of the amygdala, a nuclear group with a similar low threshold for epileptogenesis. Persistent depolarization by abnormally prolonged Na^+ currents in these neurons would favour epileptiform bursting, synchronous network activation and seizures. Alterations in *SCN5A* Na^+ channel inactivation kinetics within the limbic system are also predicted in the previously mentioned human *SCN1B* mutations with epilepsy. *SCN1B* subunits interact with *SCN5A* to markedly increase the amplitude (but not kinetics or voltage dependence) of the Na^+ current (Qu et al 1995), and mutations in the *SCN5A* gene that alter $\alpha 1$–$\beta 1$ subunit interactions also produce a variant of LQT3. Since the two subunits co-associate, the expression data therefore suggest that *SCN5A* should be considered along with *SCN1B* as a candidate Na^+ channel gene for idiopathic seizure disorders, and that epileptogenic *SCN1B* mutations may likely reveal a novel a candidate gene for human cardiac arrhythmias, although this phenotype has not yet been described in the GEFS+1 pedigree.

Ectopic localization of type II channels to hypomyelinated axons

A second major category of ion channelopathy arises from defects in neuronal regulatory pathways that control the expression, location and membrane stabilization of Na^+ channels. The molecular plasticity of Na^+ channels that follows injury to the adult axon is well described (see Waxman 2000, for review), but less is understood about inherited defects that trigger Na^+ channel rearrangements during brain development.

Molecular plasticity of Na^+ channel gene expression: the dysmyelinating shiverer mouse

One example of Na^+ channel plasticity in central pathways subsequent to a non-ion channel gene deletion is revealed by analysis of the mutant mouse shiverer, which bears a major disruption of the myelin basic protein gene, leading to absence of the glial-specific intracellular membrane protein myelin basic protein (MBP) (Roach et al 1985). Lack of MBP prevents developmental formation of the major dense line essential for the compaction of glial wrappings that form the mature myelin sheath. As a result, large calibre axons in the shiverer brain are severely hypomyelinated, with initial apposition of glial–neuronal membranes but a subsequent failure to form compact myelin and white matter in the brain.

As might be predicted from the neurological viability of the phenotype, the profound loss of myelin that should eliminate saltatory conduction is functionally rescued by striking excitability changes in the internodal axon membrane caused by up to 10-fold increases in Na^+ channel density in these otherwise inexcitable regions of axon membrane. Examination of specific 1:1 binding of [^{3}H]saxitoxin to shiverer brain revealed that the increase of Na^+ channel density was essentially confined to subcortical pathways containing hypomyelinated sensory (optic nerve), association (corpus callosum and related commissures, the centrum semiovale, and cerebellar corpus medullare), and longer-range projection fibres travelling in the internal capsule, hippocampal fimbria and fornix, while the density in grey matter regions containing the parent cell perikarya showed little increase (Fig. 2A,B). This pattern indicated that the Na^+ channel plasticity was targeted ectopically to hypomyelinated regions of central axons.

Additional analysis using subtype-specific antibodies revealed that the ectopic channel expression could be accounted for by increases in type II, but not type I or III Na^+ channels (Fig. 2B) (Westenbroeck et al 1992).

Interestingly, in the shiverer mutant, the plasticity of Na^+ channel expression appears to play both salutory and pathogenic roles. While the ectopic expression of internodal channels restores most developmental neurological function (with the exception of a residual cerebellar action tremor), shiverer mutant mice develop behavioural seizures beginning at approximately 3 months of age, culminating in premature death at 4–5 months. The seizure episodes, which occur unprovoked one or more times per day, are stereotyped and consist of sudden arrest of movement and tonic flexor spasms. The animal appears alert during the episodes, which typically last under one minute and terminate abruptly without postictal depression. The mouse promptly regains full locomotor ability, and with exception of the body tremor, appears unaffected

The mechanism of the seizure discharge in ageing shiverer mice is still unknown. However, it is tempting to speculate that continued accumulation of the ectopic Na^+ channels lowers the threshold for ephaptic cross-talk between juxtaposed axons within the large hypomyelinated fibre tracts, thus creating a basis for aberrant synchronization. However, this seizure activity apparently does not reach the level of the neocortex, since electroencephalogram (EEG) recordings obtained during these episodes display a characteristic electrodecremental pattern without the abnormal synchronous discharges typical of other epileptiform abnormalities. The shiverer lesion thus provides an example of inherited Na^+ channel-based subcortical excitability changes leading to behavioural seizures.

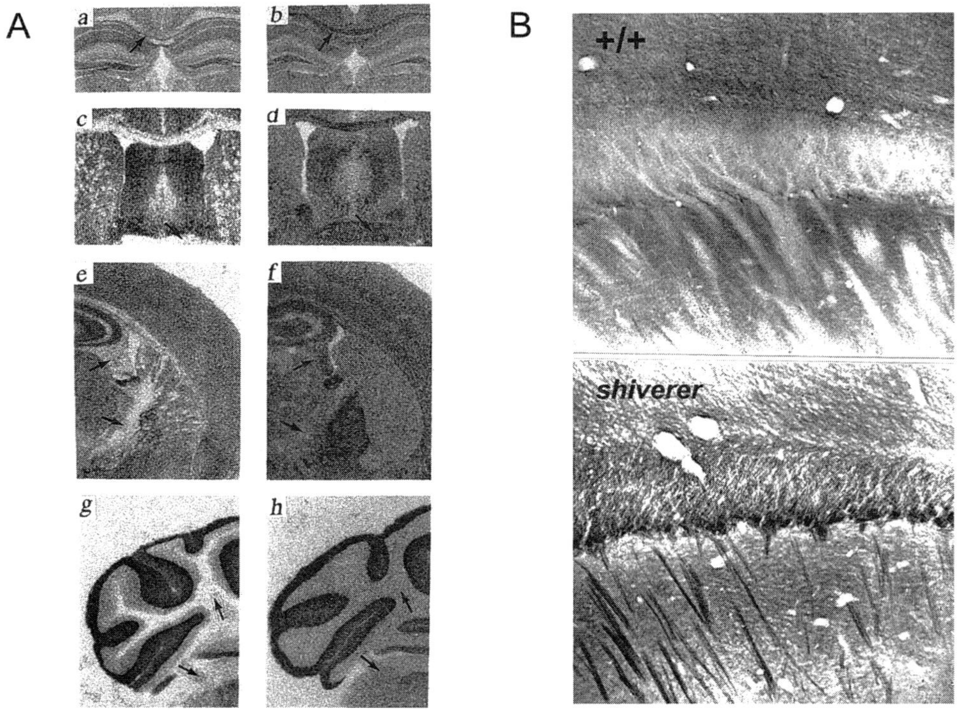

FIG. 2. (A) Selective increase of axonal Na^+ channels in hypomyelinated brain of the mutant mouse shiverer. Autoradiograms of specific [^3H]saxitoxin binding to +/+ (a,c,e,g) and shiverer (b,d,f,h) brain sections. Arrows indicate high density regions of Na^+ ion channel axon expression. (From Noebels et al 1991.) (B) Ectopic axonal Na^+ channels in shiverer brain are strongly labelled by antibody to Type II channels. Neocortex in +/+ (upper) and shiverer (lower). Note dense staining of commissural axon band (seen in cross section at cortical–subcortical junction) and descending fascicles of hypomyelinated axons in shiverer brain compared to absence of labelling in wild-type nervous system. (From Westenbroek et al 1992.)

The dysmyelinating human 18q– syndrome

The channelopathy present in the shiverer mutant may be relevant to understanding epileptogenesis in an inherited human clinical syndrome, 18q–. This mental retardation syndrome is of particular interest since the deleted chromosomal region contains the human orthologue of the myelin basic protein gene, and is accompanied by hypomyelination and characteristic seizures.

White matter changes in this neurological disorder were described only recently (Miller et al 1990, Loevner et al 1996). Both the deleted region (18q21.3–18q22.2 to the q terminus), and the clinical phenotype are highly variable, but epilepsy is present in a substantial fraction of cases, and correlates with the presence and severity of the multifocal dysmyelination (Sturm et al 2000). Video EEG monitoring of a typical seizure in an 18q– patient with epilepsy and multifocal white matter lesions revealed a brief bilateral asymmetric tonic seizure terminating in 40 seconds. The EEG showed an electrodecremental pattern within the first second, followed by generalized rhythmical α and θ rhythms with no epileptiform spiking. Interestingly, in this case, both the behavioural description and the EEG recorded during these episodes resemble, in important respects, those observed during seizure activity in the shiverer mouse model, since both show tonic contraction seizures with an absence of the high amplitude cortical discharges that typically accompany generalized tonic–clonic convulsive episodes. This pattern is consistent with a subcortical behavioural seizure origin, presumably coinciding with the hyperexcitable axons within the white matter lesions that are likely to express ectopic internodal Na^+ channels, as seen in the MBP^- shiverer mutant.

Plasticity of Na^+ channel expression in other CNS demyelinating conditions

It is worth noting that demyelinated axons studied in clinical cases of multiple sclerosis may also display continuous increases of [^{3}H]saxitoxin binding, as seen in shiverer (Moll et al 1991), indicating that similar plasticity of Na^+ channel expression can occur in some types of this heterogeneous human disorder, where it may contribute to functional neurological recovery following intermittent demyelinating attacks. In addition, multiple sclerosis is often accompanied by tonic seizures that respond to Na^+ channel blockers such as lidocaine (Sakurai & Kanazawa 1999). Thus dynamic changes in Na^+ channel biology may be part of the clinical problem, as well as part of the solution at various stages of these diseases, and excessive axonal hyperexcitability with seizures and spasms may represent the price that is paid to restore function to inexcitable fibres.

Discussion

While mutation of the Na^+ channel can probably qualify as one of the most elementary mechanisms of inherited epilepsy, and by that right should be the simplest to understand, significant complexity underlying the basis for the epileptic phenotype remains to be explored. How does one explain the considerable latency, often measured in years, from birth until seizure onset, and what are the factors that precipitate the recurring synchronous event? The induced excitability changes may also change over time. Seizures themselves up-regulate both $\alpha1$ and β Na^+ channel subunit gene expression (Bartholomei et al 1997, Gastaldi et al 1998), and this could provide yet one additional means for changes in synchronization threshold as a seizure disorder progresses. Does the location of mutant gene expression determine where the seizure actually originates, or is local connectivity more important? Do Na^+ channelopathies, like their Ca^{2+} channel counterparts (Burgess & Noebels 1999), initiate significant downstream molecular plasticity in the developing brain? Finally, how many other potential genes for epilepsy manifest their final hyperexcitability phenotype, or mask it, through downstream alterations in Na^+ channel gene expression?

Acknowledgements

Supported by NIH 29709 and the Blue Bird Circle Foundation.

References

Alekov A, Rahman MM, Mitrovic N, Lehmann-Horn F, Lerche H 2000 A sodium channel mutation causing epilepsy in man exhibits subtle defects in fast inactivation and activation *in vitro*. J Physiol 529:533–539

An RH, Wang XL, Kerem B et al 1998 Novel LQT-3 mutation affects Na^+ channel activity through interactions between α- and $\beta1$-subunits. Circ Res 83:141–146

Bartolomei F, Gastaldi M, Massacrier A, Planells R, Nicolas S, Cau P 1997 Changes in the mRNAs encoding subtypes I, II and III sodium channel alpha subunits following kainate-induced seizures in rat brain. J Neurocytol 26:667–678

Black JA, Dib-Hajj S, Cohen S, Hinson AW, Waxman SG 1998 Glial cells have heart: rH1 Na+ channel mRNA and protein in spinal cord astrocytes. Glia 23:200–208

Bland BH, Colom LV 1993 Extrinsic and intrinsic properties underlying oscillation and synchrony in limbic cortex. Prog Neurobiol 41:157–208

Burgess DL, Noebels JL 1999 Voltage-dependent calcium channel mutations in neurological disease. Ann N Y Acad Sci 868:199–212

Chen Q, Kirsch GE, Zhang D et al 1998 Genetic basis and molecular mechanism for idiopathic ventricular fibrillation. Nature 392:293–296

Escayg A, MacDonald BT, Baulac S et al 2000 Mutations of SCN1A, encoding a neuronal sodium channel, in two families with GEFS+2. Nat Genet 24:343–345

Felts PA, Yokoyama S, Dib-Hajj S, Black JA, Waxman SG 1997 Sodium channel alpha-subunit mRNAs I, II, III, NaG, Na6 and hNE (PN1): different expression patterns in developing rat nervous system. Brain Res Mol Brain Res 45:71–82

Gastaldi M, Robaglia-Schlupp A, Massacrier A, Planells R, Cau P 1998 mRNA coding for voltage-gated sodium channel β2 subunit in rat central nervous system: cellular distribution and changes following kainate-induced seizures. Neurosci Lett 249:53–56

Goldin AL 2001 Resurgence of sodium channel research. Annu Rev Physiol 63:871–894

Hartmann HA, Colom LV, Sutherland ML, Noebels JL 1999 Selective localization of cardiac SCN5A sodium channels in limbic regions of rat brain. Nat Neurosci 2:593–595

Herman LL, Stoshak M, Rittenberry TJ 1992 Long QT syndrome presenting as a seizure. Am J Emerg Med 10:435–458

Isom 2002 β subunits: players in neuronal hyperexcitability? In: Sodium channels and neuronal hyperexcitability. Wiley, Chichester (Novartis Found Symp 241) p 124–143

Kühn FJ, Greef NG 1999 Movement of voltage sensor S4 in domain 4 is tightly coupled to sodium channel fast inactivation and gating charge immobilization. J Gen Physiol 114:167–183

Loevner LA, Shapiro RM, Grossman RI, Overhauser J, Kamholz J 1996 White matter changes associated with deletions of the long arm of chromosome 18:a dysmyelinating disorder? Am J Neuroradiol 17:1843–1848

Loughney K, Kreber R, Ganetzky B 1989 Molecular analysis of the para locus, a sodium channel gene in Drosophila. Cell 58:1143–1154

Miller G, Mowrey PN, Hopper KD, Frankel CA, Ladda RL 1990 Neurologic Manifestations in 18q- syndrome. Am J Med Genet 37:128–132

Moll C, Mourre C, Lazdunski M, Ulrich J 1991 Increase of sodium channels in demyelinated lesions of multiple sclerosis. Brain Res 556:311–316

Noebels JL, Marcom PK, Jalilian-Tehrani MH 1991 Sodium channel density in hypomyelinated brain increased by myelin basic protein gene deletion. Nature 352:431–434

Otoom S, Tian LM, Alkadhi KA 1998 Veratridine-treated brain slices: a cellular model for epileptiform activity. Brain Res 789:150–156

Pacia SV, Devinsky O, Luciano DJ, Vazquez B 1994 The prolonged QT syndrome presenting as epilepsy: a report of two cases and literature review. Neurology 44:1408–1410

Pape HC, Paré D, Driesang RB 1998 Two types of intrinsic oscillations in neurons of the lateral and basolateral nuclei of the amygdala. J Neurophysiol 79:205–216

Ptacek LJ, George AL Jr, Griggs RC et al 1991 Identification of a mutation in the gene causing hyperkalemic periodic paralysis. Cell 67:1021–1027

Qu Y, Isom LL, Westenbroek RE et al 1995 Modulation of cardiac Na+ channel expression in *Xenopus* oocytes by β1 subunits. J Biol Chem 270:25696–25701

Roach A, Takahashi N, Pravtcheva D, Ruddle F, Hood L 1985 Chromosomal mapping of mouse myelin basic protein gene and structure and transcription of the partially deleted gene in shiverer mutant mice. Cell 42:149–155

Sakurai M, Kanazawa I 1999 Positive symptoms in multiple sclerosis: their treatment with sodium channel blockers, lidocaine and mexiletine. J Neurol Sci 162:162–168

Steinlein OK, Noebels JL 2000 Ion channels and epilepsy in man and mouse. Curr Opin Genet Dev 10:286–291

Sturm K, Knake S, Schomburg U et al 2000 Autonomic seizures versus syncope in 18q- deletion syndrome: a case report. Epilepsia 41:1039–1043

Wallace RH, Wang DW, Singh R et al 1998 Febrile seizures and generalized epilepsy associated with a mutation in the Na^+-channel β1 subunit gene SCN1B. Nat Genet 19:366–370

Wang Q, Shen J, Splawski I et al 1995 SCN5A mutations associated with an inherited cardiac arrhythmia, long QT syndrome. Cell 80:805–811

Wattanasirichaigoon D, Vesely MR, Duggal P et al 1999 Sodium channel abnormalities are infrequent in patients with long QT syndrome: identification of two novel SCN5A mutations. Am J Med Genet 86:470–476

Waxman SG 2000 The neuron as a dynamic electrogenic machine: modulation of sodium-channel expression as a basis for functional plasticity in neurons. Philos Trans R Soc Lond B Biol Sci 355:199–213

Westenbroek RE, Noebels JL, Catterall WA 1992 Elevated expression of type II Na^+ channels in hypomyelinated axons of shiverer mouse brain. J Neurosci 12:2259–2267

Whitaker WR, Clare JJ, Powell AJ, Chen YH, Faull RL, Emson PC 2000 Distribution of voltage-gated sodium channel alpha-subunit and beta-subunit mRNAs in human hippocampal formation, cortex, and cerebellum. J Comp Neurol 422:123–139

White JA, Alonso A, Kay AR 1993 A heart-like Na^+ current in the medial entorhinal cortex. Neuron 11:1037–1047

Yarowsky PJ, Krueger BK, Olson CE, Clevinger EC, Koos RD 1991 Brain and heart sodium channel subtype mRNA expression in rat cerebral cortex. Proc Natl Acad Sci USA 88:9453–9457

DISCUSSION

Ptacek: What do the shiverer mice look like when they are heterozygous?

Noebels: They look fine.

Ptacek: That is interesting in light of the 18q–. Presumably, the second allele is not mutated is it? Clearly, in 18q– there were a lot of other genes, but they are haploinsufficient for myelin.

Noebels: Exactly, and the dysmyelination could be a lot more severe than what we see by magnetic resonance imaging. That is an interesting point. In fact, I don't know this patient population very well. It is not clear that they all have epilepsy or severe demyelination. It is not a homogeneous syndrome. It will be interesting to find out what kind of variability there is, and whether it corresponds to heterozygous effectors.

Waxman: When you see the abnormal expression of the type 2 channel along the hypomyelinated axons in shiverer, is the type 2 channel one that has never been along those axons, or are you seeing the failure to suppress a channel that is present earlier in development?

Noebels: Good question. I think it is a failure to suppress, since type 2 is the type that is normally expressed in unmyelinated axons, for example in cerebellar granule cell axons (Westenbroek et al 1992). One question is how do these fibres know that they are not myelinated, and what tells the type 2 gene to turn on and insert channels at internodal sites that normally wouldn't be a recipient of these channels? There was a recent poster at the Society for Neuroscience meeting from Caldwell's laboratory in which immunocytochemistry was used to look at type 6 channels in shiverer (Caldwell et al 2000): they can barely see nodes, but there is some nodal assembly occurring. They also see strong expression of internodal type 2 channel isoforms. Our initial proposal was that even though there is one immediate wrap of myelin touching the axon in shiverer mice, the failure of subsequent wraps to compact, somehow the axon has to know this. Is it an inhibitory signal or an excitatory signal?

Waxman: There is currently a lot of research on how myelinated fibres build nodes. But there is a converse issue of how the previously excitable internodes are deconstructed. The yet-to-be myelinated axon has enough channels to conduct relatively securely prior to myelination and then it decreases its channel density when it is myelinated. Joel Black and I showed 15 years ago that the suppression of Na^+ channel expression in the internode depends critically on the formation of compact myelin (Black et al 1986). It doesn't occur until there is compact myelin, which makes teleological sense, otherwise we would end up with lethality. However, we don't yet know the molecular nature of that signal. It is highly focal. We also know that where there are ectopic oligodendrocyte processes touching the myelinated axon at the node, under the footprint of that process Na^+ channels disappear (Black et al 1985).

Meisler: With regard to the mechanism of the expression of SCN2A, do you know whether there is a change in the mRNA level? Or is it just a change in distribution of the proteins?

Noebels: We did one experiment with Bob Maue some years ago to examine the optic nerve, which was where we saw the largest increase in internodal SCN2A expression. I sent him shiverer tissue so that he could look at the retinal ganglion cell mRNA level to see if they were making more mRNA or not. He didn't see a difference. Before we conclude that this is just channel redistribution (which it could be), it might be that only a small increase in mRNA is necessary to repopulate axonal internodes. There is not necessarily a linear relationship between mRNA copy number and protein.

Meisler: If it were transcriptional, you ought to see a big change in whole brain.

Catterall: You don't see a big change in mRNA level in whole brain, but that is not a terribly good experiment because most of the neurons in whole brain aren't synthesizing the Na^+ channels that are in these myelinated tracts.

Goldin: In the normal situation, is SCN2A the channel at the nodes, or is it type 6?

Noebels: Type 6 is detectable at the nodes, and type 2 is not really at the nodes, except in a very small percentage of CNS fibres.

Goldin: So in those mice is there a lot of type 2 and 6 still present?

Noebels: There were $Na_v1.6$-positive nodes in shiverer, even without this myelin ever compacting. This is a little different from demyelinating lesions where the oligodendrocyte is dead. In the case of shiverer, there is an oligodendrocyte that is present but didn't make compact myelin. Hypomyelination is a very interesting lesion, compared to demyelinating diseases where oligodendrocytes die.

Ptacek: But they do have other myelin components. What happens to those?

Noebels: Then it is not called myelin. In fact, you can see that there is no white matter; the brain is all grey.

Ptacek: So the myelin that is present doesn't insulate the axon at all.

Noebels: Again, there is no myelin sheath present. Myelin only comes as a condensation of membranes as the oligodendrocyte wraps onto its compact structure. There is abundant glial membrane, but it doesn't form a myelin cylinder, and therefore it isn't seen anatomically as white matter.

Ptacek: The issue I am curious about is that part of this sounds like it is a compensation involving the up-regulation of SCN2A. But does that group of myelin proteins that are not really forming myelin still function?

Waxman: One would expect to have capacitative leak.

Noebels: We haven't yet looked at impulse conduction in these pathways, and we don't know whether this type 2 protein is even behaving as a normal type 2 channel would. It is possible that the behaviour is different. It would be nice to know whether this unmyelinated but repopulated axon is in fact behaving like a normal unmyelinated fibre or whether it is conducting better than that.

Waxman: Jeff Kocsis is doing experiments on transplantation of myelin-forming cells to the demyelinated nervous system. Remyelination can occur either by oligodendrocytes or Schwann cells. One of the things that we found is that oligodendrocytes restore the refractory period and conduction velocity to about normal levels. If we put Schwann cells in, they overcompensate and restore refractory period such that it is shorter than normal, and it is not clear whether this is due the passive properties of the Schwann cell myelin or whether there is a different signal and one is getting a difference in the Na^+ channel organization of those fibres.

Noebels: I think the latter is correct, because myelin basic protein is actually present in oligodendrocytes centrally and Schwann cells peripherally. But in the shiverer mouse, Schwann cell myelination of peripheral nerves is relatively unaffected by the same genetic lesion that prevents formation of the central white matter. The protein must either play a different role in those two cell types, or they signal to axons in different ways. Whether it is there or not, if you irritate a glial cell, the oligodendrocyte will say one thing to a central axon and the Schwann cells may say something different to a peripheral axon.

Strichartz: I have a question in a somewhat different direction. Since in brain, cardiac tissue and skeletal muscle (in T tubules) there is restricted space, K^+ accumulation is important, and K^+ hyperaccumulation during repetitive activity or prolonged depolarization may be part of a pathology. Do any of these patients show unusual sensitivities if they are put on cardiac glycosides that would exacerbate problems from extracellular K^+ accumulation?

Noebels: That is an interesting question, but I don't know the answer. High doses of these drugs tend to cause seizures by poisoning the Na^+ pump. It bears investigation. There are a number of drugs that make people with epilepsy worse, and ouabain would be one of them, particularly in these individuals.

Ptacek: I am not aware of that in muscle disease patients, but we haven't specifically asked that question.

Waxman: Astrocytes express the mRNA for several Na^+ channels, including SCN5A. Is it possible that there is a double whammy, with limbic neurons generating epileptogenesis coupled with a defect in astrocytic function that also contributes to epileptogenesis?

Noebels: I don't know. What is interesting to me is what this particular channel type is doing in these limbic circuits in the first place. I do think I understand why a mutation in SCN5A would cause epilepsy, but I don't know what this channel is doing normally.

Waxman: I'm glad that you did not ask why the channels are present in astrocytes!

Bean: What about patients with Brugada syndrome, which is caused by the other mutation in SCN5A?

Noebels: There are several mutations in SCN5A which cause idiopathic ventricular fibrillation (Chen et al 1998). It is pretty rare.

Wood: Is there no possible pharmacological intervention?

Noebels: In cases of long QT, propranolol and mexilitine are the first line of defence.

References

Black JA, Waxman SG, Hildebrand C 1985 Axo-glial relations in the retina-optic nerve junction of the adult rat: freeze–fracture observations on axon membrane structure. J Neurocytol 14:887–907

Black JA, Waxman SG, Sims TJ, Gilmore SA 1986 Effects of delayed myelination by oligodendrocytes and Schwann cells on the macromolecular structure of axonal membrane in rat spinal cord. J Neurocytol 15:745–761

Caldwell JH, Schaller KL, Lasher RS, Peles LE, Levinson SR 2000 Sodium channel Nav1.6 is localized at noses of Ranvier, dendrites and synapses. Proc Natl Acad Sci USA 97:5616–5120

Chen Q, Kirsch GE, Zhang D et al 1998 Genetic basis and molecular mechanism for idiopathic ventricular fibrillation. Nature 392:293–296

Westenbroek RE, Noebels JL, Catterall WA 1992 Elevated expression of type II Na^+ channels in hypomyelinated axons of shiverer mouse brain. J Neurosci 12:2259–2267

β subunits: players in neuronal hyperexcitability?

Lori L. Isom

Department of Pharmacology, The University of Michigan Medical School, 1301 MSRB III, Box 0632, 1150 W. Medical Center Drive, Ann Arbor, MI 48109-0632, USA

Abstract. Voltage-gated Na^+ channels are glycoprotein complexes responsible for initiation and propagation of action potentials in excitable cells such as central and peripheral neurons, cardiac and skeletal muscle myocytes, and neuroendocrine cells. Mammalian Na^+ channels are heterotrimers, composed of a central, pore-forming α subunit and two auxiliary β subunits. The α subunits form a gene family with at least 10 members. Mutations in α subunit genes have been linked to paroxysmal disorders such as epilepsy, long QT syndrome, and hyperkalaemic periodic paralysis in humans, and motor endplate disease and cerebellar ataxia in mice. Three genes encode Na^+ channel β subunits with at least one alternative splice product. A mutation in the β1 subunit gene has been linked to generalized epilepsy with febrile seizures plus type 1 (GEFS+1) in a human family with this disease. Na^+ channel β subunits are multifunctional. They modulate channel gating and regulate the level of channel expression at the plasma membrane. More recently, they have been shown to function as cell adhesion molecules in terms of interaction with extracellular matrix, regulation of cell migration, cellular aggregation, and interaction with the cytoskeleton. Structure–function studies have resulted in the preliminary assignment of functional domains in the β1 subunit. A Na^+ channel signalling complex is proposed that involves β subunits as channel modulators as well as cell adhesion molecules, other cell adhesion molecules such as neurofascin and contactin, RPTPβ, and extracellular matrix molecules such as tenascin.

2002 Sodium channels and neuronal hyperexcitability. Wiley, Chichester (Novartis Foundation Symposium 241) p 124–143

β subunits modulate Na^+ channel gating

Voltage-gated Na^+ channels isolated from mammalian central and peripheral neurons and cardiac myocytes are heterotrimeric complexes composed of α, β1, and β2 subunits (Catterall 1992, Malhotra et al 2001). α and β1 subunits have also been identified in skeletal muscle (Trimmer et al 1989, Yang et al 1993) while β2 has not. The α subunits encode a small gene family of at least 10 members (Plummer & Meisler 1999). Recently, two additional members of the β subunit gene family, β1A (a splice variant of SCN1B) (Kazen-Gillespie et al 2000) and $\beta3$ (SCN3B) (Morgan et al 2000), have been described. α subunit cDNAs express functional Na^+ channels in heterologous expression systems such as *Xenopus* oocytes or transfected mammalian fibroblasts. However, at least for brain and skeletal muscle α subunits, the properties of these channels expressed in isolation are very different from native currents. Coexpression of the auxiliary β subunits with these channels results in significant shifts in the voltage-dependence of activation and inactivation, changes in channel modal gating behaviour resulting in increases in the rate of inactivation as well as the rate of recovery from inactivation (Isom et al 1994), and increases in channel density at the plasma membrane as assessed by saxitoxin binding studies (Isom et al 1995b).

Na^+ channel gating and the effect of β subunit coexpression are dependent on the particular heterologous expression system used. For example, brain and skeletal muscle α subunits gate slowly when expressed in *Xenopus* oocytes. Addition of $\beta1$ results in major changes in channel gating mode (Isom et al 1994). In contrast, these same channels expressed in mammalian fibroblasts gate in a fast mode in the absence of β subunits and β subunit coexpression has more subtle effects than those observed in oocytes, suggesting that the major function of β subunits in neurons may be something in addition to kinetic modulation of Na^+ channels (Isom et al 1995b). Interestingly, not all Na^+ channel α subunits are detectably modulated by $\beta1$ and $\beta2$ subunits, especially those channels primarily expressed in peripheral nerve (Sangameswaran et al 1996, Souslova et al 1997). Perhaps these channels do not associate with β subunits or require yet unidentified auxiliary subunits.

Recent studies have shown that tsA201 cells and HEK-293 cells, frequently used as heterologous expression systems for voltage-gated Na^+ channels, express endogenous $\beta1$ and $\beta1A$ subunits, respectively (Malhotra et al 2001, Moran et al 2000). The levels of endogenous β subunit expression in these lines appear to be passage-dependent and variable between sub-lines (Isom laboratory, unpublished results). These findings may necessitate the reinterpretation of some of the Na^+ channel expression data in the literature. Furthermore, future investigators must exercise caution to define the level of endogenous β subunit expression in cell lines prior to transfection with Na^+ channel α subunits.

Na^+ channel β subunits, the *Drosophila* Na^+ channel auxiliary subunit TipE, and the β subunits of voltage-gated K^+ and Ca^{2+} channels have similar functions in terms of modulation of channel activity (Adelman 1995, Isom et al 1994, Warmke et al 1997). As will be discussed below, however, only the mammalian Na^+ channel β subunits also function as cell adhesion molecules (CAMs). What other proteins associate with Na^+ channels and contribute to the ion conduction proteome in excitable cells? How is Na^+ channel function affected by these associated proteins and by the local cellular environment? Recent advances, outlined below, have begun to answer these important questions in channel biology.

Na^+ channel β subunits are structurally homologous to cell adhesion molecules

β1, β1A, β2 and β3 are transmembrane proteins with type I topology: containing an extracellular N-terminus, a single transmembrane segment, and an intracellular C-terminus. β1 and β1A are splice variants of the *SCN1B* gene (Kazen-Gillespie et al 2000). β3 is closely related to β1 and is the product of a separate gene, *SCN3B* (Morgan et al 2000). The expression of *SCN1B* and *SCN3B* mRNA is developmentally and anatomically regulated. *In situ* hybridization studies comparing β1 and β3 localization in the adult CNS suggested that the expression of these two subunits is complementary (Morgan et al 2000). The developmental expression patterns of β1 and β1A RNA showed that β1A is expressed early in embryonic brain development while β1 is not (Kazen-Gillespie et al 2000). β1A expression decreases after birth, concomitant with the emergence of β1 expression. In the adult rat, immunolocalization studies showed β1A protein expression in heart and dorsal root ganglion neurons with limited expression in brain (Kazen-Gillespie et al 2000).

All four β subunit molecules contain extracellular immunoglobulin (Ig) domains that are structurally homologous to the V-set of the Ig superfamily that includes CAMs (Isom & Catterall 1996). This unique property of the Na^+ channel auxiliary subunits was first discovered following sequence analysis of β2, revealing that its extracellular domain contained an Ig fold and an extended region with similarity to the CAM contactin (Isom et al 1995a). Two distinct regions of the extracellular domain of contactin have greater than 40% amino acid sequence identity with Na^+ channel β2 subunits. Subsequent analysis of the extracellular domains of β1/β1A and β3 showed a similar homology to the CAM myelin P_o (McCormick et al 1998, Morgan et al 2000). Investigation of the amino acid sequences of β1, β1A, β2 and β3 shows that these proteins are closely related (Isom 2001). The Ig loop region (Isom et al 1994) of all four proteins is well conserved. While β3 is most closely related to β1, there are a number of conserved residues between β3 and β2 as well.

The Ig loop region of β1 is essential for α subunit interactions

The extracellular domain of β1 is critical for modulation of channel gating and voltage-dependence of brain and skeletal muscle Na^+ channels expressed in *Xenopus* oocytes (Chen & Cannon 1995, Makita et al 1996, McCormick et al 1998, 1999). Removal of the intracellular domain of β1, substitution of the intracellular β1 domain with the corresponding domains from either β2 or the structurally related CAM myelin P_o, or attachment of the β1 extracellular domain to the plasma membrane via a glycosylphosphatidylinositol anchor all resulted in retention of full β1-like modulation of Nav1.2 channels expressed in *Xenopus* oocytes (McCormick et al 1998). Mutagenesis of the extracellular $\beta1$ Ig loop, based on comparison with the crystal structure of the Ig loop of myelin P_0, resulted in identification of key residues in the A/A' face of $\beta1$ required for interaction with $Na_v1.2$ in oocytes (McCormick et al 1998). In contrast, mutagenesis of hydrophobic residues at the juxtamembrane region of $\beta1$ or the predicted sites of *N*-linked glycosylation had no effect on $\beta1$-mediated channel modulation in oocytes (McCormick et al 1998). A mutation in the putative transmembrane region did alter the voltage-dependence of steady state inactivation, suggesting that this region may also play a role in α–$\beta1$ interactions (McCormick et al 1998).

Where do α and $\beta1$ interact? The studies described above suggested strongly that the extracellular Ig domain of $\beta1$ subunits interacts with α subunits through a site formed by one or more extracellular loops of the α subunit. Experiments were designed to identify domains critical for α–$\beta1$ subunit interactions using brain, skeletal muscle, and cardiac Na^+ channel α subunit cDNAs expressed in *Xenopus* oocytes (Makita et al 1996). The design of these experiments took advantage of the observation that brain ($Na_v1.2$) and skeletal muscle ($Na_v1.4$) Na^+ channels are modulated from slow to fast gating modes by coexpression of $\beta1$ in *Xenopus* oocytes. In contrast, cardiac Na^+ channels ($Na_v1.5$) gate in a fast mode in the absence of $\beta1$ subunits when expressed in oocytes. A series of $Na_v1.4$/$Na_v1.5$ chimeras expressed in oocytes showed that substitution of IS5–S6 and IVS5–S6 in $Na_v1.4$ with the corresponding domains in $Na_v1.5$ completely abolished the effects of $\beta1$ on channel inactivation (Makita et al 1996). Substitution of just one of these domains resulted in channels with an attenuated response to $\beta1$, suggesting that $\beta1$ may interact with a discontinuous epitope formed by IS5–S6 and IVS5–S6 that are in close proximity to each other in the 3D structure of the Na^+ channel. Alternatively, $\beta1$ may interact elsewhere on the channel, but these domains are required for changes in gating behaviour. More recently, using a series of $Na_v1.2$/$Na_v1.5$ chimeras it was shown that substitution of the IVSS2–S6 segment in $Na_v1.2$ with the corresponding domain in $Na_v1.5$ produced a channel that did not shift to a fast gating mode in the presence of $\beta1$ subunits. (Qu et al 1999) (IVSS2–S6 is contained within the IVS5–S6 region used in the first study.) This reduced $\beta1$ effect could be partially overcome by the addition of higher levels of $\beta1$ subunits in the oocyte. Thus, it was suggested that the IVSS2–S6 substitution resulted in a reduction in the affinity of the α subunit for $\beta1$. It may also be possible that substitution of an additional domain in the α subunit, for example, IS5–S6 as shown in the first study, is required to completely abolish the effects of $\beta1$.

Are α–$\beta1$ interactions strictly extracellular? As discussed below, a mutation in the C-terminal domain of $Na_v1.5$, D1790G, has been implicated in LQT-3 (An et al 1998). According to topology predictions, D1790 is located intracellularly. This mutation has been shown to result in the disruption of α–$\beta1$ subunit interactions when expressed in transfected HEK cells. Other studies have shown that removal of the C-terminal domain of $\beta1$ subunits results in a protein that retains activity in *Xenopus* oocytes, as shown previously, but does not fully modify Na^+ channels when expressed in transfected mammalian cells (Malhotra et al 1999, Meadows et al 2001). These studies suggest that α and $\beta1$ may have both intracellular and extracellular interaction domains and/or that channel modulation is more complex in mammalian cells than in *Xenopus* oocytes, perhaps requiring other signalling molecules.

β subunits function as CAMs

Do Na^+ channel β subunits play roles in addition to modulation of channel kinetics? Two key observations provided important clues. First, $\beta2$ subunit expression was found to be required to detect the translocation of newly synthesized Na^+ channels from intracellular stores to the plasma membrane of primary cultured embryonic neurons (Schmidt et al 1985, Schmidt & Catterall 1986). Second, expression of $\beta2$ subunits in the absence of α subunits in *Xenopus* ooctyes resulted in dramatic changes in membrane capacitance through promotion of intracellular vesicle fusion with the microvillus membrane (Isom et al 1995a). $\beta1/\beta1A$, $\beta2$, and $\beta3$ are structurally related to CAMs. We and others asked whether β subunits actually function as CAMs using a number of criteria (Malhotra et al 2000, Srinivasan et al 1998, Xiao et al 1999). Do they interact with the extracellular matrix? Do they interact with other cell adhesion molecules leading to cellular aggregation and cytoskeletal changes? Do they participate in cell migration? If so, what are the implications of these activities on Na^+ channel localization and density at the node of Ranvier or specialized domains in other excitable cells?

Na^+ channel β subunits interact with extracellular matrix proteins and influence cell migration

Glial-derived extracellular matrix molecules, for example tenascin C (TN-C) and tenascin R (TN-R), play important roles in cellular interactions in developing or injured neuronal cells, for example in migration, neuritogenesis and regeneration (Faissner 1997, Schachner 1997). The tenascins are multi-functional molecules that can promote neurite outgrowth, inhibit growth cone advance, and induce axonal defasciculation *in vitro*. Na^+ channel β subunits interact with TN-C (34) and TN-R (Srinivasan et al 1998, Xiao et al 1999). Incubation of purified Na^+ channels with TN-C showed saturable and specific binding (Srinivasan et al 1998). Glutathione *S*-transferase (GST) fusion proteins containing various domains of TN-C and TN-R were tested for their ability to bind purified Na^+ channels or the recombinant $\beta2$

subunit extracellular domain. Both Na^+ channels and $\beta 2$ bound specifically to the fibronectin (FN) type III repeats 1–2, A, B and 6–8 of TN-C and FN type III repeats 1–2 and 6–8 of TN-R. Transfected cells expressing $\beta 1$ or $\beta 2$ were repelled from TN-R plated on a nitrocellulose substrate (Xiao et al 1999). The same TN-R GST fusion proteins used in the first study were used to determine which domains were responsible for the observed repulsion. Both $\beta 1$- and $\beta 2$-expressing cell lines were strongly repelled by EGF-L (cysteine-rich amino terminus of TN-R plus the EGF-like domains), but adhered well to EGF-S (EGF-like domains only), FN 6-8, FG and GST, suggesting that the N-terminus of TN-R may be involved in repulsion of $\beta 1$- or $\beta 2$-expressing cells. Cells expressing $\beta 1$ subunits alone initially adhered to the TN-R recombinant domains FN 6-8 (as found in the first study for $\beta 2$) and EGF-S prior to repulsion. A mixture of EGF-L, EGF-S, and FN6-8 fusion proteins added to the cell culture medium blocked the adhesion of $\beta 1$-expressing cells to the EGF-like or fibronectin-like domains of TN-R in a concentration-dependent manner. β subunit-mediated effects in response to TN-R occurred in the absence of α subunits, suggesting that β subunits may function as CAMs independently of the ion channel complex. In two-electrode recordings in *Xenopus* oocytes, EGF-L fusion protein produced a rapid increase in the amplitude of Na^+ currents (Xiao et al 1999). EGF-L-mediated current potentiation was observed in oocytes expressing $Na_v1.2$ α subunits alone and in oocytes coexpressing $Na_v1.2$ α, $\beta 1$ and $\beta 2$ subunits. In contrast, neither FN 6-8 fusion protein nor GST affected Na^+ currents in oocytes, suggesting that current potentiation is a specific effect of the EGF-L domain of TN-R. EGF-L-mediated potentiation was not accompanied by any detectable changes in the voltage-dependence of current activation or inactivation or in any obvious effects on current time course.

Na^+ channel β subunits interact homophilically to cause cellular aggregation and recruitment of ankyrin to the plasma membrane

CAMs of the Ig superfamily interact homophilically and heterophilically to transduce signals between adjacent cells or adjacent axons where they participate in, for example, axonal fasciculation, growth cone guidance, and nodal formation. L1-like CAMs (a subgroup of the Ig superfamily) interact homophilically in a *trans* mechanism to induce cellular aggregation (Hortsch & Bieber 1991, Kamiguchi & Lemmon 1997). Following homophilic binding, L1 CAMs that have intracellular C-terminal domains transduce signals resulting in the recruitment of ankyrin and spectrin to points of cell–cell contact. Using the *Drosophila* S2 cell model system we tested whether Na^+ channel $\beta 1$ and $\beta 2$ subunits behave in a similar manner. S2 cells transfected with $\beta 1$, $\beta 2$, $\alpha\beta 1$, or $\alpha\beta 2$ subunits displayed homophilic interactions while mock transfected cells or cells transfected with $Na_v1.2$ α alone did not

(Malhotra et al 2000). Immunocytochemical analysis of the cell aggregates revealed recruitment of endogenous ankyrin to sites of cell–cell contact, indicating that extracellular β subunit-mediated adhesion transduces a signal through the plasma membrane resulting in cytoskeletal communication.The intracellular domains of $\beta 1$ and $\beta 2$ are critical for ankyrin recruitment (Malhotra et al 2000). Mutant constructs lacking the intracellular C-terminal domains of $\beta 1$ and $\beta 2$, respectively, were transfected into S2 cells and the cells were induced to aggregate. While both cell lines aggregated, ankyrin staining was diffuse and not concentrated to points of cell–cell contact as in cells expressing full length β subunits. We concluded that, while the extracellular Ig domains are sufficient to mediate cellular aggregation, the intracellular C-terminal domains of $\beta 1$ and $\beta 2$ are required for homophilic adhesion-mediated signal transduction to the cytoskeleton. A conserved phosphotyrosine residue in the cytoplasmic domain of L1 CAMs has been shown to be critical for ankyrin recruitment (Garver et al 1997). Interestingly, $\beta 1$, $\beta 1$A and $\beta 3$ each contain a single tyrosine residue in their intracellular domains. These intracellular tyrosine residues may play important roles in cell adhesion-mediated ankyrin recruitment in the $\beta 1$-like auxiliary subunits of Na^+ channels. Association of $\beta 1$ subunits with receptor tyrosine phosphatase β (RPTPβ) (Ratcliffe et al 2000) may regulate the phosphorylation state of the tyrosine residue and thus modulate Na^+ channel–cytoskeletal interactions. Cytoskeletal interactions through ankyrin and the β subunits may be critical to Na^+ channel placement in excitable cells. In *Xenopus* oocytes, $Na_v1.4$ and $Na_v1.8$ Na^+ channels expressed in the absence of $\beta 1$ subunits shift to a fast gating mode when sites of membrane attachment to the cytoskeleton are ruptured by patch excision, negative pressure or agents that disrupt microtubule formation (Shcherbatko et al 1999). Application of positive pressure prevented this shift in kinetics. The authors suggested that because $\beta 1$ subunits cause similar changes in Na^+ channel gating in oocytes, $\beta 1$ may also function through cytoskeletal interactions. In other studies, treatment of inside-out patches of ventricular cells with cytochalasin-D induced $Na_v1.5$ Na^+ channels to enter a mode characterized by lower peak open probability with a greater persistent activity, consistent with a decrease in the rate of inactivation (Undrovinas et al 1995). Cardiac Na^+ channels in ankyrin$_B$ knockout mice display reduced Na^+ current density and abnormal kinetics that contribute to prolonged action potential duration and abnormal QT-rate adaptation (Chauhan et al 2000). Thus, cytoskeletal interactions may be critical to Na^+ channel localization and gating in excitable tissues such as heart and brain.

Characterization of β subunits in cardiac myocytes

What is the function of Na^+ channel β subunits in excitable cells other than neurons? We recently completed a study characterizing Na^+ channel α and β

subunits in cardiac myocytes and showed that $\beta 1$ and $\beta 2$ are expressed and associated with $Na_v1.1$ and $Na_v1.5$ α subunits. Immunocytochemical experiments with specific antibodies showed that α ($Na_v1.1$ and $Na_v1.5$), $\beta 1$ and $\beta 2$ subunits are localized to z lines in heart sections (Cohen 1996, Malhotra et al 2001). Thus, as in neurons, Na^+ channels in cardiac myocytes are localized to discrete functional domains.

A number of hypotheses have been proposed regarding the physiological role of β subunits in heart. Brain ($Na_v1.1$, $Na_v1.2$, $Na_v1.3$ and $Na_v1.6$) and skeletal muscle ($Na_v1.4$) Na^+ channel α subunits expressed in oocytes exhibit slow inactivation kinetics (Catterall 2000). Coexpression of $\beta 1$ subunits produces a significant increase in the rate of inactivation of these channels (Catterall 2000). In contrast, expression of $Na_v1.5$ α subunits in oocytes produces channels that inactivate rapidly in the absence of β subunits (Qu et al 1995). Some groups have reported that $\beta 1$ has no observable effects on $Na_v1.5$ functional expression (Makita et al 1994, Yang et al 1993). Others have reported that coexpression of $\beta 1$ and $Na_v1.5$ α subunits results in increased Na^+ current density with no detectable effects on channel kinetics or voltage-dependence (Nuss et al 1995, Qu et al 1995). Some groups have found modulation of channel sensitivity to lidocaine block and subtle changes in channel kinetics and gating properties in response to $\beta 1$ expression (Makielski et al 1996), while others have reported significant shifts in the voltage-dependence of steady-state inactivation and activation, as well as recovery from inactivation (An et al 1998, Malhotra et al 2001, Xiao et al 2000). Coexpression of $Na_v1.5$ and $\beta 1$ subunits resulted in modification of fatty acid blockade of Na^+ current (Xiao et al 2000). Expression of $\beta 1$ and $\beta 2$ subunits was required to observe robust slip-mode conductance in transfected cardiac Na^+ channels (Nuss et al 1999). A $Na_v1.5\alpha$ subunit mutation associated with long-QT syndrome affects the voltage dependence of channel inactivation by altering the interaction of $Na_v1.5$ α and $\beta 1$ subunits (An et al 1998). Finally, $Na_v1.1$ α subunits are modulated by $\beta 1$ and $\beta 2$ subunits when expressed in oocytes (Smith & Goldin 1998). Thus, Na^+ channel β subunits may modulate cardiac Na^+ channels and play a role in cardiac as well as neuronal physiology.

Na^+ channel β subunits are involved in human disease

GEFS+ epilepsy. Mutations in voltage-gated Na^+ channel α subunit genes have been shown to be linked to inherited human diseases causing paroxysmal events like long QTsyndrome (LQT) in the heart and hyperkalaemic periodic paralysis in skeletal muscle (Ackerman & Clapham 1997). Until recently there were no identified neuronal Na^+ channel mutations associated with human disease. Mutations in *SCN1A* have been linked to two families with idiopathic generalized epilepsy with febrile seizures plus type 2 (GEFS+2) (Escayg et al 2000). A mutation in

SCN1B has been shown to be linked to a large family with GEFS+1 (Wallace et al 1998). This mutation, C121W (or C102W as originally defined; Isom et al 1992), changes a critical cysteine residue defining the C-terminal region of the Ig loop in $\beta 1$ to a tryptophan, presumably destabilizing this critical cell adhesion and α subunit association domain. Coexpression of the mutant $\beta 1$ subunit with rat brain $Na_v1.2$ channels in *Xenopus* oocytes revealed a lack of the classic $\beta 1$-mediated acceleration of the inactivation rate (Wallace et al 1998). The expressed current was virtually identical to that observed with α alone. It is not yet known whether C121W mutant $\beta 1$ subunits are expressed at the plasma membrane or, if so, whether they are associated with α subunits. It will be important in future studies to examine the coexpression of mutant C121W $\beta 1$ subunits and α subunits in transfected mammalian cells as well as in transgenic animals. Nevertheless, this important study showed for the first time that mutations in Na^+ channel auxiliary subunits cause diseases of the human nervous system.

Long QT syndrome. LQT, an inherited cardiac arrhythmia that includes prolonged ventricular repolarization, has been shown to involve mutations in Na^+ and K^+ channel genes. LQT-3 has been attributed to mutations in *SCN5A*, often resulting in channels that fail to inactivate completely during prolonged depolarization. A novel LQT-3 mutation in *SCN5A*, D1790G, has been shown to affect the interaction of α and $\beta 1$ subunits (An et al 1998). Expression of mutant D1790G $Na_v1.5$ alone in HEK 293 cells resulted in channels with normal biophysical properties as compared to wild-type, monomeric $Na_v1.5$ channels. However, channels formed by the coexpression of mutant $Na_v1.5$ and $\beta 1$ were markedly different from wild-type. Unlike previously reported LQT-3 mutations, channels formed by D1790G $Na_v1.5$ and $\beta 1$ did not conduct sustained inward current with prolonged depolarization. In contrast, the steady state inactivation of D1790G $Na_v1.5$ channels was shifted by −16 mV. The authors concluded that the effect of this mutation on inactivation of Na^+ channels was due to D1790G-induced changes in α–$\beta 1$ subunit interactions.

Brugada syndrome. Brugada syndrome, a form of idiopathic ventricular fibrillation (IVF) characterized by right bundle-branch block with ST-segment elevation, has been shown to be due to mutations in *SCN5A* (Chen et al 1998). Coexpression of one of these mutant $Na_v1.5$ subunits (R1232W+T1620M) with $\beta 1$ subunits in HEK cells resulted in a marked reduction in Na^+ current amplitude compared to expression of the mutant α subunit alone (Wan et al 2000). The authors hypothesized that IVF mutations of *SCN5A* reduce the number of functional Na^+ channels via a $\beta 1$-dependent mechanism and thus reduce the maximum available Na^+ conductance in the heart. Interestingly, the effect of $\beta 1$ subunits on (R1232W+T1620M) mutant channels was partially prevented at a lower temperature (26 °C vs. 37 °C). Another Brugada syndrome mutant $Na_v1.5$ (T1620M) exhibits subtle shifts in the voltage-dependence of inactivation and acceleration of recovery from inactivation when expressed alone in *Xenopus* oocytes. Coexpression with $\beta1$ enhanced both of these effects, resulting in a further positive shift in the voltage-dependence of inactivation and further acceleration of the rate of recovery from inactivation. The authors suggested that $\beta1$ expression causes a severe functional defect by exposing a large overlap in the relationship between activation and inactivation, or a window current, inducing hyperexcitability and arrhythmias (Makita et al 2000).

Neuropathic pain. In rat models of neuropathic pain, chronic constrictive injury results in dynamic changes in the relative expression of Na^+ channel α subunits in the dorsal root ganglion as well as in the spinal cord. In the Bennett and Xie model, levels of $\beta1$ and $\beta2$ subunit mRNA in the dorsal horn of the spinal cord are differentially regulated (Blackburn-Munro & Fleetwood-Walker 1999). $\beta1$ mRNA levels increased, while $\beta2$ mRNA levels decreased significantly within laminae I-II on the ipsilateral side of the spinal cord relative to the contralateral side. In laminae II-IV $\beta1$ mRNA levels remained constant while $\beta2$ levels again showed a small but significant decrease. The authors proposed that a functional down-regulation of $\beta2$ subunits may decrease the interaction of β subunits with tenascins, thereby promoting axonal growth and projection of neurons into the superficial laminae where new synaptic contacts could be initiated. It remains to be seen whether β subunit protein levels correspond to the reported changes in mRNA following neuronal injury. Changes in β subunit protein expression would be expected to regulate Na^+ channel density and localization in the neuronal plasma membrane or modulate interactions between Na^+ channels and the cytoskeleton or extracellular matrix.

Conclusions

Na^+ channel β subunits play important roles in channel modulation and regulation of channel density at the plasma membrane. β subunits also function as cell adhesion molecules in heterologous expression systems. What are the potential implications of these findings? It has been shown previously that $ankyrin_G$, the CAMs neurofascin and NrCAM, and voltage-gated Na^+ channels colocalize at the axonal membrane of the adult node of Ranvier in specialized membrane domains (Davis et al 1996, Kordeli et al 1990). Early clusters of neurofascin and NrCAM are joined later by $ankyrin_G$ and Na^+ channels during differentiation of myelinated axons. Formation of the node of Ranvier may then result from the fusion of two cluster intermediates. An $ankyrin_G$-mediated link between neurofascin, NrCAM, and ion channels may allow these CAMs to cluster Na^+

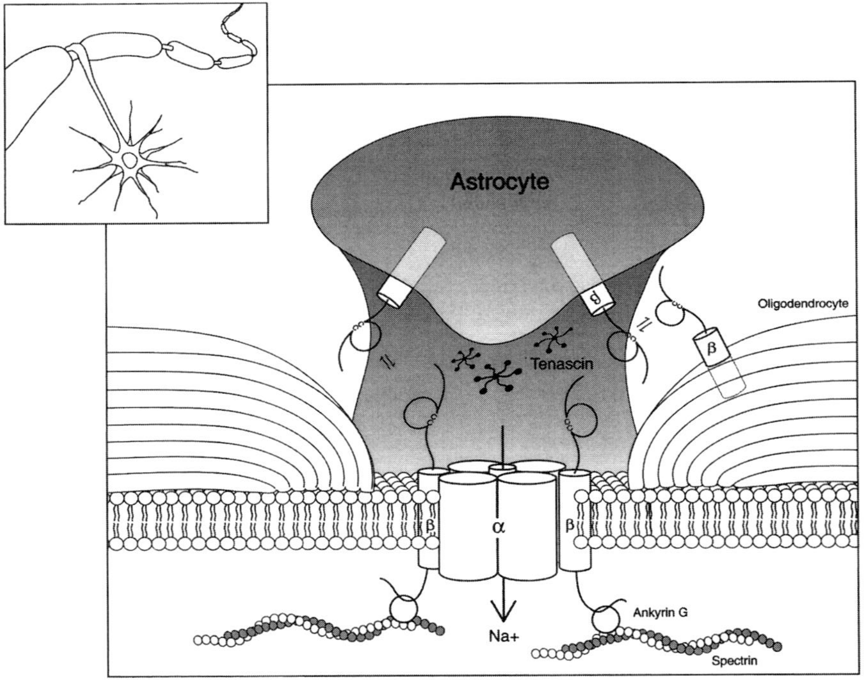

FIG. 1. Model of voltage-gated Na^+ channel β subunit cell adhesive interactions at the node of Ranvier. Na^+ channel β subunits are cell adhesion molecules of the Ig superfamily. In the nervous system, they are expressed in neurons, including the node of Ranvier, and in glia. β subunits modulate Na^+ channel gating and plasma membrane expression levels, interact with the extracellular matrix protein TN-R, and communicate with other β subunits through *trans* homophilic cell adhesive interactions to recruit ankyrin to the plasma membrane. β subunits may also communicate with other CAMs at the node of Ranvier such as neurofascin, NrCAM, or contactin via *cis* heterophilic interactions or other β subunits in the node via *cis* homophilic adhesion. We propose that β subunits on astrocyte projections, on the axon and on oligodendrocytes or Schwann cells communicate to modulate Na^+ channel density and localization at nodes of Ranvier. (*Insert*) Astrocyte projection encircling node of Ranvier on a myelinated axon. (From Isom 2001, with permission.)

channels in the axonal membrane. It was proposed that Na^+ channel $\beta 2$ subunits, because of their homology to F3/contactin, may interact in a lateral or *cis* fashion with NrCAM and thus contribute to Na^+ channel localization. Our data expand this model to propose a direct link between Na^+ channels and $ankyrin_G$ through homophilic β subunit interactions (Malhotra et al 2000) (Fig. 1). The multivalent membrane binding domain of $ankyrin_G$ (Michaely & Bennett 1993) may allow interaction with multiple CAMs, including neurofascin, NrCAM, as well as Na^+ channel β subunits, forming a Na^+ channel signalling complex at the node of Ranvier that may also include contactin (Kazarinova-Noyes et al 2001). β subunit mRNA expression has been described in sciatic nerve Schwann cells, astrocytes from spinal cord, optic nerve and sciatic nerve, and B104 oligodendrocyte precursor cells in culture (Dib-Hajj et al 1996, Oh et al 1997, Oh & Waxman 1994). *Trans*-homophilic cell adhesion may occur between axonal and glial cell Na^+ channel β subunits. This putative adhesion may also contribute to Na^+ channel clustering in nodes of Ranvier during the process of myelination.

It has been proposed that cardiac Na^+ channels may be targeted and clustered to specific locations in a similar manner to that observed for Na^+ channels in brain (Cohen 1996). The presence of β subunits in cardiac myocytes may facilitate Na^+ channel localization and clustering to discrete functional domains via homophilic or heterophilic cell adhesion interactions. Interestingly, treatment of inside-out patches of ventricular cells with cytochalasin-D induced Na^+ channels to enter a mode characterized by lower peak open probability with a greater persistent activity, consistent with a decrease in the rate of inactivation (Undrovinas et al 1995). Na^+ channels in $ankyrin_B$ knockout mice display reduced current density and abnormal kinetics that contribute to prolonged action potential duration and abnormal QT-rate adaptation (Chauhan et al 2000). Thus, cytoskeletal interactions may be critical to Na^+ channel localization and gating in the heart as well as in the brain. We propose that the presence of $\beta 1$ and $\beta 2$ subunits in cardiac myocytes may facilitate Na^+ channel–cytoskeletal interactions and play a key role in the regulation of the cardiac action potential. Our challenge now is to relate these exciting observations in heterologous expression systems to physiological events.

Acknowledgements

This work was supported by grants from the National Institutes of Health (1R01MH59980-01A1) and the National Science Foundation (IBN-9734462) to LLI.

References

Ackerman M J, Clapham DE 1997 Ion channels — basic science and clinical disease. N Eng J Med 336:1575–1586 (erratum: 1997 N Engl J Med 337:579)

Adelman JP 1995 Proteins that interact with the pore-forming subunits of voltage-gated ion channels. Curr Opin Neurobiol 5:286–295

An RH, Wang XL, Kerem B et al 1998 Novel LQT-3 mutation affects Na^+ channel activity through interactions between alpha- and beta1-subunits. Circ Res 83:141–146

Blackburn-Munro G, Fleetwood-Walker SM 1999 The sodium channel auxiliary subunits $\beta 1$ and $\beta 2$ are differentially expressed in the spinal cord of neuropathic rats. Neuroscience 90:153–164

Catterall WA 1992 Cellular and molecular biology of voltage-gated sodium channels. Physiol Rev 72:S15–S48

Catterall WA 2000 From ionic currents to molecular mechanisms: the structure and function of voltage-gated sodium channels. Neuron 26:13–25

Chauhan VS, Tuvia S, Buhusi M, Bennett V, Grant AO 2000 Abnormal cardiac Na^+ channel properties and QT heart rate adaptation in neonatal $ankyrin_B$ knockout mice. Circ Res 86:441–447

Chen C, Cannon SC 1995 Modulation of Na^+ channel inactivation by the $\beta 1$ subunit: a deletion analysis. Eur J Physiol 431:186–195

Chen Q, Kirsch GE, Zhang D et al 1998 Genetic basis and molecular mechanism for idiopathic ventricular fibrillation. Nature 392:293–296

Cohen SA 1996 Immunocytochemical localization of rH1 sodium channel in adult rat heart atria and ventricle. Presence in terminal intercalated disks. Circulation 94:3083–3086

Davis JQ, Lambert S, Bennett V 1996 Molecular composition of the node of Ranvier: identification of ankyrin- binding cell adhesion molecules neurofascin (mucin+/third FNIII domain−) and NrCAM at nodal axon segments. J Cell Biol 135:1355–1367

Dib-Hajj SD, Hinson AW, Black JA, Waxman SG 1996 Sodium channel mRNA in the B104 neuroblastoma cell line. FEBS Lett 384:78–82

Escayg A, MacDonald BT, Meisler MH et al 2000 Mutations of SCN1A, encoding a neuronal sodium channel, in two families with GEFS+2. Nat Genet 24:343–345

Faissner A 1997 The tenascin gene family in axon growth and guidance. Cell Tissue Res 290:331–341

Garver TD, Ren Q, Tuvia S, Bennett V 1997 Tyrosine phosphorylation at a site highly conserved in the L1 family of cell adhesion molecules abolishes ankyrin binding and increases lateral mobility of neurofascin. J Cell Biol 137:703–714

Hortsch M, Bieber AJ 1991 Sticky molecules in not-so-sticky cells. Trends Biochem Sci 16:283–287

Isom LL 2002 Sodium channel beta subunits: anything but auxiliary. Neuroscientist 7:42–54

Isom LL, Catterall WA 1996 Na^+ channel subunits and Ig domains. Nature 383:307–308

Isom LL, De Jongh KS, Patton DE et al 1992 Primary structure and functional expression of the $\beta 1$ subunit of the rat brain sodium channel. Science 256:839–842

Isom LL, De Jongh KS, Catterall WA 1994 Auxiliary subunits of voltage-gated ion channels. Neuron 12:1183–1194

Isom LL, Ragsdale DS, De Jongh KS et al 1995a Structure and function of the $\beta 2$ subunit of brain sodium channels, a transmembrane glycoprotein with a CAM motif. Cell 83:433–442

Isom LL, Scheuer T, Brownstein AB, Ragsdale DS, Murphy BJ, Catterall WA 1995b Functional co-expression of the beta 1 and type IIA alpha subunits of sodium channels in a mammalian cell line. J Biol Chem 270:3306–3312

Kamiguchi H, Lemmon V 1997 Neural cell adhesion molecule L1: signalling pathways and growth cone motility. J Neurosci Res 49:1–8

Kazarinova-Noyes K, Malhotra JD, McEwen DP et al 2001 Contactin associates with Na^+ channels and increases their functional expression. J Neurosci 21:7517–7525

Kazen-Gillespie KA, Ragsdale DS, D'Andrea MR, Mattei LN, Rogers KE, Isom, LL 2000 Cloning, localization, and functional expression of sodium channel $\beta 1A$ subunits. J Biol Chem 275:1079–1088

Kordeli E, Davis J, Trapp B, Bennett V 1990 An isoform of ankyrin is localized at nodes of Ranvier in myelinated axons of central and peripheral nerves. J Cell Biol 110:1341–1352

Makielski JC, Limberis JT, Chang SY, Fan Z, Kyle JW 1996 Coexpression of beta 1 with cardiac sodium channel alpha subunits in oocytes decreases lidocaine block. Mol Pharmacol 49:30–39

Makita N, Bennett PB Jr, George AL Jr 1994 Voltage-gated Na^+ channel $\beta 1$ subunit mRNA expressed in adult human skeletal muscle, heart, and brain is encoded by a single gene. J Biol Chem 269:7571–7578

Makita N, Bennett PB, George AL Jr 1996 Molecular determinants of $\beta 1$ subunit-induced gating modulation in voltage-dependant Na^+ channels. J Neurosci 16:7117–7127

Makita N, Shirai N, Wang D et al 2000 Cardiac Na^+ channel dysfunction in Brugada syndrome is aggravated by $\beta 1$-subunit. Circulation 101:54–60

Malhotra JD, Meadows L, Ragsdale DS, Hortsch M, Isom LL 1999 Cytoplasmic domains of voltage-gated sodium channel β subunits are essential for ankyrin recruitment and channel modulation in heterologous expression systems. Soc Neurosci Abstr 25:1731

Malhotra JD, Kazen-Gillespie K, Hortsch M, Isom LL 2000 Sodium channel β subunits mediate homophilic cell adhesion and recruit ankyrin to points of cell–cell contact. J Biol Chem 275:11383–11388

Malhotra JD, Chen C, Rivolta I et al 2001 Characterization of sodium channel α and β subunits in rat and mouse cardiac myocytes. Circulation 103:1303–1310

McCormick KA, Isom LL, Ragsdale D, Smith D, Scheuer T, Catterall WA 1998 Molecular determinants of Na^+ channel function in the extracellular domain of the beta1 subunit. J Biol Chem 273:3954–3962

McCormick KA, Srinivasan J, White K, Scheuer T, Catterall WA 1999 The extracellular domain of the $\beta 1$ subunit is both necessary and sufficient for $\beta 1$-like modulation of sodium channel gating. J Biol Chem 274:32638–32646

Meadows L, Malhotra JD, Stetzer A, Isom LL, Ragsdale DS 2001 The intracellular segment of the sodium channel $\beta 1$ subunit is required for efficient modulation with the sodium channels alpha subunit. J Neurochem 76:1871–1878

Michaely P, Bennett V 1993 The membrane-binding domain of ankyrin contains four independently folded subdomains, each comprised of six ankyrin repeats. J Biol Chem 268:22703–22709

Moran O, Nizzari M, Conti F 2000 Endogenous expression of the $\beta 1$A sodium channel subunit in HEK-293 cells. FEBS Lett 473:132–134

Morgan K, Stevens EB, Shah B et al 2000 $\beta 3$: An additional auxiliary subunit of the voltage-sensitive sodium channel that modulates channel gating with distinct kinetics. Proc Natl Acad Sci USA 97:2308–2313

Nuss HB, Chiamvimonvat N, Pérez-García MT, Tomaselli GF, Marbán E 1995 Functional association of the beta 1 subunit with human cardiac (hH1) and rat skeletal muscle ($\mu 1$) sodium channel alpha subunits expressed in Xenopus oocytes. J Gen Physiol 106:1171–1191

Nuss HB, Marbán E, Balke CW 1999 Whether 'slip-mode conductance' occurs. Science 284:711

Oh Y, Waxman SG 1994 The $\beta 1$ subunit mRNA of the rat brain Na^+ channel is expressed in glial cells. Proc Natl Acad Sci USA 91:9985–9989

Oh Y, Lee YJ, Waxman SG 1997 Regulation of Na^+ channel $\beta 1$ and $\beta 2$ subunit mRNA levels in cultured rat astrocytes. Neurosci Lett 234:107–110

Plummer NW, Meisler MH 1999 Evolution and diversity of mammalian sodium channel genes. Genomics 57:323–331

Qu Y, Isom LL, Westenbroek RE et al 1995 Modulation of cardiac Na^+ channel expression in *Xenopus* oocytes by $\beta 1$ subunits. J Biol Chem 270:25696–25701

Qu Y, Rogers JC, Chen SF, McCormick KA, Scheuer T, Catterall WA 1999 Functional roles of the extracellular segments of the sodium channel alpha subunit in voltage-dependent gating and modulation by beta 1 subunits. J Biol Chem 274:32647–32654

Ratcliffe CF, Qu Y, McCormick KA et al 2000 A sodium channel signaling complex: modulation by associated receptor protein tyrosine phosphatase beta. Nat Neurosci 3:437–444
Sangameswaran L, Delgado SG, Fish LM et al 1996 Structure and function of a novel voltage-gated, tetrodotoxin-resistant sodium channel specific to sensory neurons. J Biol Chem 271:5953–5956
Schachner M 1997 Neural recognition molecules and synaptic plasticity. Curr Opin Cell Biol 9:627–634
Schmidt JW, Catterall WA 1986 Biosynthesis and processing of the alpha subunit of the voltage-sensitive sodium channel in rat brain neurons. Cell 46:437–445
Schmidt JW, Rossie S, Catterall WA 1985 A large intracellular pool of inactive Na channel alpha subunits in developing rat brain. Proc Natl Acad Sci USA 82:4847–4851
Shcherbatko A, Ono F, Mandel G, Brehm P 1999 Voltage-dependent sodium channel function is regulated through membrane mechanics. Biophys J 77:1945–1959
Smith RD, Goldin AL 1998 Functional analysis of the rat I sodium channel in *Xenopus* oocytes. J Neurosci 18:811–820
Souslova VA, Fox M, Wood JN, Akopian AN 1997 Cloning and characterization of a mouse sensory neuron tetrodotoxin-resistant voltage-gated sodium channel gene, Scn10a. Genomics 41:201–209
Srinivasan J, Schachner M, Catterall WA 1998 Interaction of voltage-gated sodium channels with the extracellular matrix molecules tenascin-C and tenascin-R. Proc Natl Acad Sci USA 95:15753–15757
Trimmer JS, Cooperman SS, Tomiko SA et al 1989 Primary structure and functional expression of a mammalian skeletal muscle sodium channel. Neuron 3:33–49
Undrovinas AI, Shander GS, Makielski JC 1995 Cytoskeleton modulates gating of voltage-dependent sodium channel in heart. Am J Physiol 269:H203–H214
Wallace RH, Wang DW, Singh R et al 1998 Febrile seizures and generalized epilepsy associated with a mutation in the Na^+-channel $\beta 1$ subunit gene SCN1B. Nat Genet 19:366–370
Wan X, Wang Q, Kirsch GE 2000 Functional suppression of sodium channels by $\beta 1$-subunits as a molecular mechanism of idiopathic ventricular fibrillation. J Mol Cell Cardiol 32:1873–1884
Warmke JW, Reenan RA, Wang P et al 1997 Functional expression of Drosophila para sodium channels. Modulation by the membrane protein TipE and toxin pharmacology. J Gen Physiol 110:119–133
Xiao ZC, Ragsdale DS, Malhorta JD et al 1999 Tenascin-R is a functional modulator of sodium channel beta subunits. J Biol Chem 274:26511–26517
Xiao YF, Wright SN, Wang GK, Maorgan JP, Leaf A 2000 Coexpression with beta(1)-subunit modifies the kinetics and fatty acid block of hH1(alpha) Na^+ channel. Am J Physiol Heart Circ Physiol 279:H35–H46
Yang JS, Bennett PB, Makita N, George AL, Barchi RL 1993 Expression of the sodium channel $\beta 1$ subunit in rat skeletal muscle is selectively associated with the tetrodotoxin-sensitive α subunit isoform. Neuron 11:915–922

DISCUSSION

Noebels: Isn't the recovery in the peripheral nervous system preventing that astrocyte–node interaction? In this case, wouldn't you expect to see some differences between PNS and CNS?

Isom: Yes, and we do. For instance, in our work with Peter Shrager (Kazarinova-Noyes et al 2001) where we see the interaction of Na^+ channels with contactin, contactin is colocalized to adult optic nerve nodes of Ranvier, but not sciatic nerves. The cell adhesion interactions appear to be very different in those two areas.

Noebels: Have you introduced the GEFS1+ mutation that is supposed to disrupt that extracellular protein fold, and does this affect cell adhesion?

Isom: Yes, in S2 cells β1C121W does not cause aggregation, although it is expressed (Loukas et al 2001). In mammalian cells and oocytes we can see modulation of the channel kinetics, but we have to put in a high concentration of RNA. I think it is an affinity issue, so we can push the association if we add a lot (Loukas et al 2001). In the original paper (Wallace et al 1998) they showed no modulation of the channel, and I think this is true at that concentration. If you put in a high concentration, you can make it act like β1.

Segal: When we record Na^+ currents using outside–out patch recordings, we get relatively long channel openings. When we record cell-attached patch recordings we get much shorter openings, a difference noted by Aldrich & Stevens (1987). Our natural inclination is to think that the cell-attached patch recordings are more physiological. One hypothesis to explain the difference between different patch types is that there are different cytoskeletal interactions, but it seems that we could imagine an effect directly at the β subunits explaining this. Do you think this is plausible? Do you have any suggestions as to how we might coax the inside–out patches into a more physiological state?

Isom: Once you interrupt interactions with the underlying cytoskeleton, the characteristics of the current change. Have you seen Shcherbatko's paper on the same subject in oocytes (Shcherbatko et al 1999)? A shift to fast inactivation could be accomplished by patch excision. They predicted that β1 may interact with the cytoskeleton.

Horn: Our results suggest that this probably won't be to do with the cytoskeleton. We see the same thing with a cell-attached patch and an inside–out patch. The cell-attached patch itself disrupts the cytoskeleton when you draw the membrane into the patch pipette. I think something else is going on when you pull it away. There is probably still some cytoskeleton in there, and I think the membrane is disrupted in inside–out patches.

Segal: We tested a few cytoskeletal hypotheses and came up with no method of making the recordings physiological (M. Segal & A. Douglas, unpublished results).

Horn: Also, I think a cell-attached patch is already unphysiological. There were some early reports from Buzz Brown's lab that when you make a cell-attached patch, within a couple of minutes there are some shifts in the gating behaviour, as if something were already changing.

Isom: That shift happens over about the first 10 min.

Ptacek: It could be phosphorylation, for example. That is one experiment you could do *in vitro*.

Catterall: I agree that you should think about phosphorylation, and not only serine and threonine phosphorylation but also tyrosine phosphorylation. In relatively recent experiments, we found that a receptor tyrosine phosphatase is associated with Na^+ channels, and dephosphorylates the α subunit with a functional effect on the inactivation. You need to consider not only protein kinase A and protein kinase C, but also tyrosine kinases and phosphatases whose activity may be changed when you go from an intracellular milieu to either a cell-attached or inside–out patch. Adding ATP is probably not enough to reconstitute that regulation, because it is not clear that you have all the kinases that would be part of the complex when you pull a patch off.

Segal: It seems to be an important methodological issue to figure out, not just for its own sake, but also because we will probably be seeing mutations in such regulatory systems operating on these channels.

Catterall: The tyrosine phosphatase interacts with both α and $\beta 1$ subunits. So far we have only been able to show that it dephosphorylates α, but I fully expect that we will find conditions under which it dephosphorylates the tyrosine residue that Lori Isom described on the $\beta 1$ subunit (Malhotra et al 2001).

I wanted to come back to the point that Lori Isom made about the different modulation of Na^+ channel function by the β subunits that was observed in her original experiments on oocytes, compared to what is observed in the knockout mouse experiments. In oocytes the voltage dependence of steady-state inactivation is shifted to the left by co-expression of $\beta 1$ subunits. This was also the case when we made continuous cell lines expressing α and $\beta 1$ in the CHO cell background, but there was a difference between the oocytes and the cell lines in that the inactivation was slow in the oocytes but fast in the CHO cell lines. More recently, we have done the same experiments in transiently transfected HEK 293 cells. When we coexpress β subunits with Na^+ channel α subunits in HEK cells, we get shifts of the voltage dependence of inactivation in the opposite direction, positive relative to α subunits alone. This result correlates with the mouse experiment in which knockout of $\beta 2$ causes a shift to the left in the voltage dependence of inactivation. The conclusion is that a mouse hippocampal neuron is more like an HEK cell than it is like a CHO cell or a *Xenopus* oocyte. I think it means that the effects of β subunits are dependent on cell-type-specific signal transduction, and no absolute conclusion can be made. The effect depends on the modulatory effects of cell signalling pathways before you put the $\beta 1$ subunit in.

Waxman: Following up on that, how finely parsed is the nervous system in terms of channels manifesting their presence differently in different types of neurons? Bruce Bean, I think you have made the point that the type 6 channel expressed in a CA3 neuron is different from the type 6 channel expressed in a Purkinje cell.

Catterall: We only know the tip of the iceberg, because so few classes of neurons are studied in this regard. Everyone studies hippocampal or cortical neurons except for a few people like Bruce who study Purkinje cells.

Bean: Along similar lines, do you have any speculation about the mechanism by which the tenascin R piece is regulating the Na^+ current? Do you think this is by affecting the interaction with the β subunits?

Isom: Yes. You can see modulation in the absence of β subunits. My student has recently shown by gel overlay and co-immunoprecipitation that tenascin-R binds directly to the α subunit as well as β (Davis et al 2001). They all bind, but they bind different domains of tenascin R. The β subunits bind similar domains and the α subunit binds a different domain.

Bean: Are these extracellular domains?

Isom: Yes.

Bean: Is there any relation to the scorpion toxin binding sites?

Isom: We haven't looked that closely yet.

Strichartz: What are the indications that the α and either of the two β subunits associate in Golgi before they get to the plasma membrane?

Isom: That is a critical experiment that we need to do. The closest anyone has got is John Schmidt, in Bill Catterall's lab, who showed that during Na^+ channel biosynthesis, association with $\beta 2$ was the very final step. There is a large intracellular pool of free α subunits. When they are at the cell membrane, then they are associated with $\beta 2$, so we think it is a late step.

Strichartz: But you think that the α subunits would be membrane associated, so if they are intracellular, they are likely to be found on Golgi. The β subunits are also membrane associated so they would also be in Golgi, but there is no indication that there is a complex formed at the level of Golgi.

Isom: We don't know that yet.

Horn: Didn't Bill Agnew show that there is an association in the plasma membrane in oocytes? He showed that you could express α alone and then inject the next day an antisense α along with a β subunit, and they would associate in the plasma membrane.

Isom: Yes.

Horn: Is that believable? Could the association between α and β in the plasma membrane be dynamic?

Isom: I think it is, and our mutagenesis studies are starting to suggest this. We have a new paper coming out in which we have looked at a mutant called $\beta 1$ stop that lacks the intracellular tail (Meadows et al 2001). In oocytes, this mutant associates very poorly with α and has approximately a 10 000-fold lower affinity for α than wild-type $\beta 1$. It is expressed as efficiently as wild-type $\beta 1$ yet does not associate with α efficiently. In transfected cells only a small portion of the total mutant $\beta 1$ pool can be co-immunoprecipitated with α, even though they are both present in the plasma membrane.

Ptacek: The Tyr181Glu mutant caused a phenotype of not concentrating at the nodes. Do you think that phosphorylation keeps it away from the nodes?

Isom: Yes.

Ptacek: One could synthesize a phosphotyrosine-specific antibody. Then it would be interesting to look at the shiverer mouse, to look for differences between phosphotyrosine $\beta1$ versus the non-phosphorylated form.

Isom: Yes. The reason we went after that is because in the L1 family of cell adhesion molecules an intracellular tyrosine residue is critical for cell adhesion and ankyrin recruitment. We predict that the tyrosine in $\beta1$ will be very important in formation of the Na^+ channel signalling complex (Malhotra et al 2001).

Ptacek: So you think that phosphorylation of $\beta1$ might lead to redistribution of Na^+ channels?

Isom: Yes, that is where Bill Catterall's paper showing the interaction of $\beta1$ with RPTP β comes in. RPTP β may regulate whether $\beta1$ is communicating with ankyrin.

Noebels: I wanted to follow up on Bill Catterall's earlier comment on the diversity of action. Do you think it is paradoxical that the $\beta2$ knockout reduced the Na^+ population by 30%, but it has a lower seizure threshold and is a hyperexcitable nervous system? (Bharucha et al 2000).

Isom: There could be a reduction in Na^+ channels in inhibitory neurons.

Noebels: So you are suggesting that if we look carefully at where these channels are lost from, the phenotype may make more sense?

Segal: The other possibility is that the persistent Na^+ current is the most important determinant of excitability. The excitability increase from a doubling of the persistent Na^+ current could outweigh a 30% decrease in channel number.

Catterall: In the hippocampal neurons there is an obvious change in persistent Na^+ current in the $\beta2$ knockouts.

Goldin: Does the decreased number of channels mean that the safety factor is decreased, so as you raise temperature, you begin to see the effect?

Isom: That may be true. We need to challenge these mice.

Ptacek: I presume you are going to breed the double knockouts.

Isom: Yes, we have started that. They will probably live to about day 10.

Baker: Could the Na^+ channel expression in Schwann cells and astrocytes be related to their ability to stick to axons?

Isom: Yes. One thing we are looking for is whether or not β subunits are expressed anywhere by themselves without α subunits. This will be pretty easy to see for $\beta2$. We have seen $\beta1$ expressed in tissues like epithelia where you wouldn't expect to find Na^+ channels. I wonder whether they can be expressed as cell adhesion molecules in their own right without the channel. That is what I think they would be doing in the glia.

Strichartz: With respect to the question about whether you get hyperexcitability or positive neurological signs with a deficit of Na^+ channels, I think there are several neurological examples where there are actually deficits that lead to positive signs. Post-herpetic neuralgia is one of them. If you take skin biopsies from people who are suffering from shingles, they have a paucity of nerve endings and reduced innervation in that area, yet they are extraordinarily allodynic. The arguments are vague, but one of them is that if you partially lack the same afferent input into the dorsal horn, you get inappropriate sprouting of intact, primary sensory neurons. Perhaps if you have less than normal activity that is present in the nervous system you will get an inappropriate expression of other afferent fibres there or even have altered descending influences. You don't necessarily need more input or more channel activity, or more channels, in order to get hyperexcitability or positive signs.

Cummins: Herpes can reduce Na^+ channel density in those neurons. That is one of the effects that has been reported.

One of the problems with the sensory neuronal channels is getting them functionally expressed. There is a feeling that there is some β subunit or other factor that we are missing.

Isom: Whatever it is, it is not homologous to anything we have. We have screened a thousand times and nothing else falls out. We are co-immunoprecipitating Na^+ channels out of target tissues and then doing Western blotting to identify what we can with the antibodies that we have available, and then we ask what the unidentified proteins are.

Cummins: Do you see $\beta 1$, $\beta 2$ and $\beta 3$?

Isom: Yes.

References

Aldrich RW, Stevens CF 1987 Voltage-dependent gating of single sodium channels from mammalian neuroblastoma cells. J Neurosci 7:418–431

Bharucha VA, Chen C, Chen Y et al 2000 Mice lacking sodium channel $\beta 2$ subunits have reduced sodium channel densities and negative shifts in the voltage dependence of channel inactivation. Soc Neurosci Abstr 26:1109

Davis T, Buchberger J, Malhotra JD et al 2001 Association between the α and $\beta 1$ subunits of neuronal sodium channels and tenascin-R. Soc Neurosci Abstr 27

Kazarinova-Noyes K, Malhotra JD, McEwen DP et al 2001 Contactin associates with Na^+ channels and increases their functional expression. J Neurosci 21:7517–7525

Loukas A, Kriegler S, Kazen-Gillespie KA et al 2001 Dose-dependent modulation and suppression of sodium channel currents by mutant $\beta 1$ subunit associated with GEFS+1 epilepsy. Soc Neurosci Abstr 27

Malhotra JD, Koopman MC, Kazen-Gillespie KA, Hortsch M, Isom LL 2001 Modulation of sodium channel $\beta 1$ subunit-mediated ankyrin recruitment by an intracellular tyrosine residue. Soc Neurosci Abstr 27

Meadows L, Malhotra JD, Stetzer A, Isom LL, Ragsdale DS 2001 The intracellular segment of the sodium channel $\beta 1$ subunit is required for efficient association with the sodium channels alpha subunit. J Neurochem 76:1871–1878

Shcherbatko A, Ono F, Mandel G, Brehm P 1999 Voltage-dependent sodium channel function is regulated through membrane mechanics. Biophys J 77:1945–1959

Wallace RH, Wang DW, Singh R et al 1998 Febrile seizures and generalized epilepsy associated with a mutation in the Na^+-channel beta 1 subunit gene SCN1B. Nat Genet 19:366–370

Modulation of sodium channels in primary afferent neurons

Stuart Bevan and Nina Storey

Novartis Institute for Medical Sciences, 5 Gower Place, London WC1E 6BN, UK

Abstract. Electrophysiological studies have revealed that the properties of voltage-gated Na^+ channels can be modified by phosphorylation. Na^+ channels have multiple sites for phosphorylation by protein kinases A and C (PKA and PKC). A change in the phosphorylation state of Na^+ channels is an important mechanism of neuromodulation for both central and peripheral neurons. In isolated primary afferent sensory neurons, application of an inflammatory mediator, prostaglandin E_2 (PGE_2), causes an increase in excitability associated with a hyperpolarizing shift in the activation curve of the tetrodotoxin-resistant (TTX-R) Na^+ currents. The experimental evidence indicates that the effect of PGE_2 is mediated by an elevation in cAMP levels and activation of PKA. This potentiation of TTX-R Na^+ channel activity is in marked contrast to the inhibitory effects of PKA and PKC on tetrodotoxin-sensitive (TTX-S) currents in central neurons. Infection of dorsal root ganglion neurons with *Herpes simplex* virus (HSV) results in an abolition of excitability associated with a selective loss of both TTX-S and TTX-R Na^+ currents: voltage-gated Ca^{2+} and K^+ channels are unaffected by HSV infection. The loss of Na^+ current is due to a virally induced internalization process and requires extracellular Na^+.

2002 Sodium channels and neuronal hyperexcitability. Wiley, Chichester (Novartis Foundation Symposium 241) p 144–158

Control of sensory neuron excitability by modulation of Na^+ channels

Primary afferent neurons transduce and transmit sensory signals (including temperature, pressure, proprioception and pain) from the peripheral tissues and organs to the CNS. Sensory signalling in mammalian afferent neurons has been particularly well studied for nociception where a range of inflammatory mediators, such as prostaglandins, can sensitize the neurons and increase their ability to respond to noxious chemical, mechanical and thermal stimuli. Recent studies have revealed that modification of voltage-gated tetrodotoxin-resistant (TTX-R) Na^+ channel properties can contribute to the sensitization process by increasing the probability that a depolarizing stimulus will evoke action potentials in the sensory neurons. This modification of TTX-R channel properties involves protein kinases A and C (PKA and PKC) mediated phosphorylation events and is in marked contrast to the effects of phopshorylation of tetrodotoxin-sensitive (TTX-S) channels in CNS neurons which are associated with current inhibition and a consequent reduction in excitability.

Phosphorylation of Na^+ channels

There is ample biochemical evidence that the α subunits of brain Na^+ channels can be phosphorylated on multiple sites following activation of PKA and PKC (Costa & Catterall 1984a,b). Conversely Na^+ channels can be dephoshorylated by phosphatases, notably by calcineurin and phosphatase 2A (Murphy et al 1993, Chen et al 1995), suggesting that Na^+ channels are dynamically regulated by the degree of phosphorylation. In general, electrophysiological studies of either native or heterologously expressed cloned TTX-S Na^+ channels (notably $Na_v1.2$) have shown that both PKC (Numann et al 1991) and PKA (Li et al 1992, Smith & Goldin 1996) activation lead to a reduction in peak macroscopic Na^+ conductance, usually with no shift in the voltage dependences of activation and steady state availability. PKC activation also leads to a slowing of inactivation, which is associated with longer single channel openings and persistent channel activity with prolonged depolarizations (Numann et al 1991). These modifications explain, in part, the ability of agents that activate PKA to reduce the excitability of CNS neurons.

Site-directed mutagenesis studies have identified the major sites for phosphorylation-mediated alterations in TTX-S Na^+ channel activity. The PKA-induced reduction in peak conductance can be ascribed to phosphorylation of serine 571 in the I–II intracellular linker region of the channel (Cantrell et al 1997, Smith & Goldin 1996), while the PKC-induced reduction is due to phosphorylation of serine residues 554 and 573 (Cantrell et al 2000). Mutagenesis studies have also indicated that phosphorylation of serine 1506 is responsible for the PKC-mediated slowing of inactivation (West et al 1992). The PKC and PKA phosphorylation events for TTX-S channels are not independent and PKC phosphorylation is, at least in some experimental conditions, required for the PKA mediated reduction in Na^+ conductance (Li et al 1993).

Excitatory effects of PGE_2 on DRG neurons are mediated by cAMP and PKA

Application of prostaglandin E_2 (PGE_2) to dorsal root ganglion (DRG) neurons isolated from either adult or neonatal rats leads to an increase in the number of action potentials evoked by depolarizing stimuli (England et al 1996, Fig. 1). This effect can be ascribed, in a large part, to an effect on the TTX-R Na^+ channels (England et al 1996, Gold et al 1996) that are a feature of these sensory

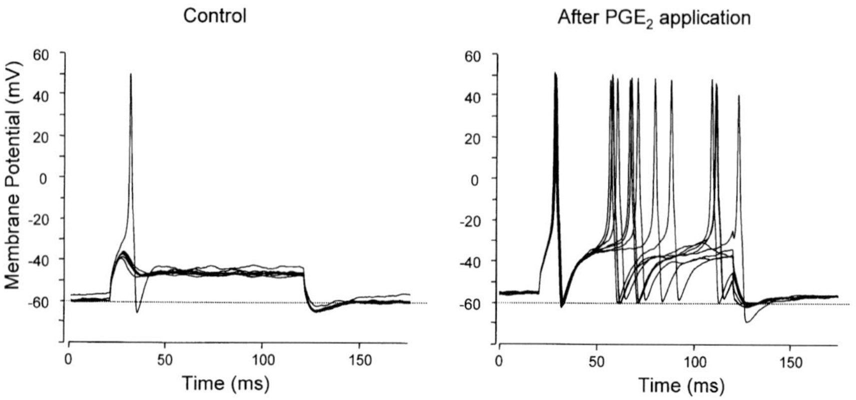

FIG. 1. Responses of DRG neuron to current injection in presence of 500 nM TTX. (Left) 200 pA current pulses evoke no action potentials until 1 M PGE_2 added when a single action potential is evoked. (Right) 30 s later same current pulses evoke multiple action potentials. Data from England et al (1996).

neurons, as the facilitatory effects of PGE_2 are seen when the TTX-S currents are blocked by high concentrations of TTX. Two types of TTX-R channels have been identified in DRG neurons ($Na_v1.8$ and $Na_v1.9$); however, it is probable that the reported effects of hyperalgesic agents on TTX-R currents will have reflected actions on $Na_v1.8$ under the experimental conditions used.

In voltage clamp experiments, application of PGE_2 to DRG neurons increased the amplitude of TTX-R currents evoked by depolarizing steps to test potentials of −10 mV. This increase in amplitude occurred over several minutes and was associated with increases in the rates of activation and decay of the currents. Figure 2 shows that the basis of these changes was a ~5 mV shift in the voltage dependence of activation to more hyperpolarized potentials (V_{50} for activation shifted from −4.5 mV to −10.1 mV). A similar hyperpolarizing shift in the steady state inactivation relationship (V_{50} −32.3 to −39.3 mV) was noted after PGE_2 treatment (England et al 1996). Although the peak current seen in the current–voltage relationship was increased by about 13% after PGE_2 application, this was primarily due to the change in driving force with the hyperpolarizing shift in activation and the reported overall conductance was variable with both increases and reductions reported (England et al 1996, Gold et al 1996).

The PGE_2-activated prostanoid receptors on DRG neurons (EP_1, EP_4, and some splice variants of EP_3 receptors—EP_{3B} and EP_{3C}) can be coupled to the stimulation of adenylate cyclase to raise cAMP levels. Experimental elevation of cAMP with the adenylate cyclase stimulator, forskolin, or application of membrane permeant cAMP analogues, dibutyryl cAMP (dbcAMP) and 8-bromo-cAMP, mimicked the effect of PGE_2 treatment by shifting the activation

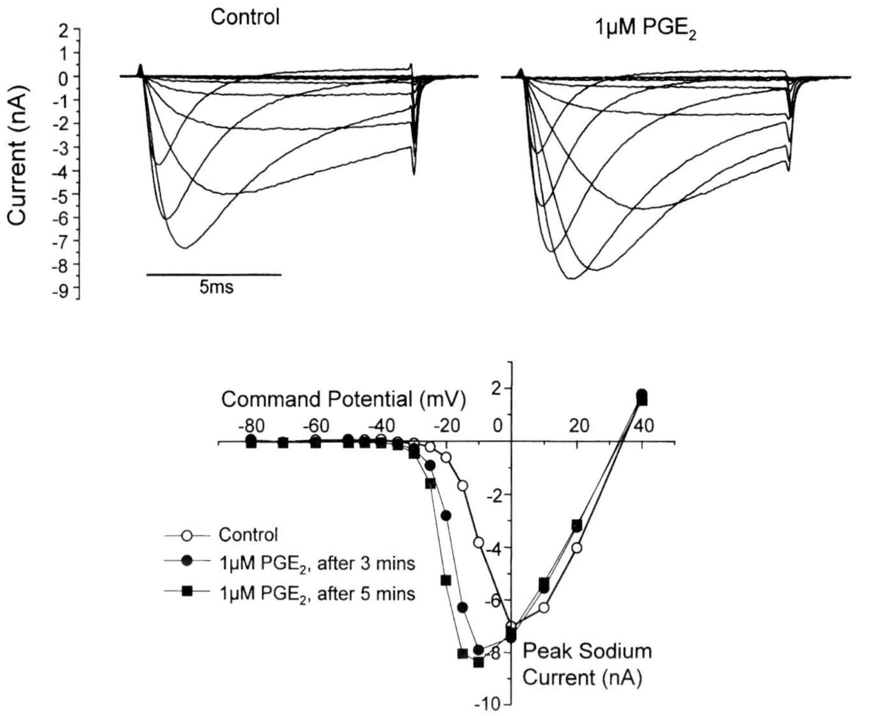

FIG. 2. (Top) Families of TTX-R currents evoked under control conditions (left) and after exposure to PGE_2. (Bottom) Peak current–voltage relationships under control conditions and 3 and 5 minutes after PGE_2 addition. Data from England et al (1996).

and steady-state inactivation curves in the hyperpolarizing direction (England et al 1996, Gold et al 1996, Fig. 3). In contrast no such effects were noted with the forskolin analogue, dideoxyforkolin, which does not activate adenylate cyclase. In fact dideoxyforskolin decreased the peak conductance without evoking any shift in the current–voltage relationship for activation (England et al 1996). Prior exposure to dbcAMP also abrogated the effects of subsequent PGE_2 applications on TTX-R currents, which argues for a cAMP-mediated pathway for the PGE_2 effect. This cAMP pathway appears to involve PKA activation as the effects of PGE_2 were inhibited by intracellular application of peptide inhibitors of PKA (England et al 1996, Gold et al 1996).

PKC activity regulates TTX-R currents

PKC regulation of TTX-R currents in DRG neurons was first shown by application of staurosporine or intracellular administration of the peptide

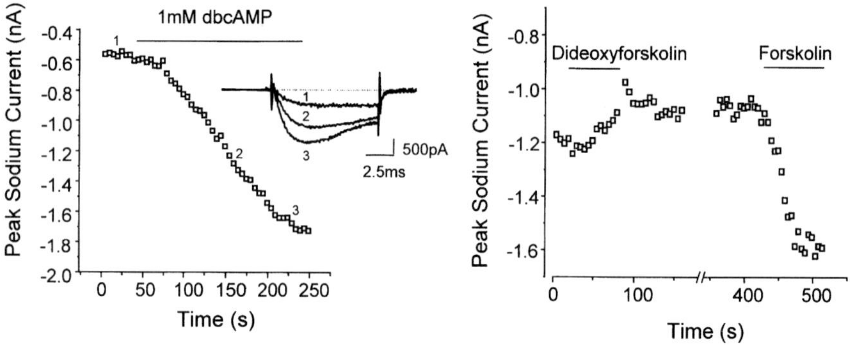

FIG. 3. (Left) 1 mM dibutyryl cAMP evokes an increase in TTX-R current evoked by depolarizations to a test potential of −15 mV. (Inset) Currents evoked at times 1,2,3 indicated on graph. (Right) Dideoxyforskolin (10 M) evokes a small reduction in TTX-R current amplitude while a subsequent challenge with forskolin (10 M) increases current amplitude. Data from England et al (1996).

inhibitor PKC_{19-36} (Gold et al 1998). These treatments decreased peak current without any significant effect on the voltage dependence of activation. Conversely, the PKC activators phorbol-12-myristate,13-acetate (PMA) and phorbol 12,13-dibutyrate (PDBu) dose-dependently increased the amplitudes of TTX-R currents in DRG neurons by a mechanism that could be inhibited by the peptide inhibitor PKC_{19-36}. However, although the inhibitory effects of the two PKC stimulators were blocked by this specific inhibitor, there were some differences between the effects of these two agents. The effects of PMA were not associated with any shift in the time course or voltage dependence of activation, whereas PDBu induced changes in the rates of current activation and inactivation without any shift in the voltage dependence of activation. This reason for the different actions of the two PKC activators is unclear but it has been interpreted to suggest activation of different PKC isoforms.

PKA and PGE_2 modulation of TTX-R currents requires PKC activity

The presence of an interaction between PKC and PKA in modulating TTX-R currents was demonstrated by the finding that the PKC inhibitors, PKC_{19-36} and staurosporine, significantly attenuated the ability of forskolin to modulate TTX-R currents. In contrast, the PKA inhibitors WIPTIDE and Rp-cAMPs failed to affect the PDBu-mediated changes in increases in peak TTX-R current conductance, although the PDBu effects on kinetics were reduced (Gold et al 1998).

PKC inhibitors were also able to attenuate the effects of PGE_2 on TTX-R currents in DRG neurons and prior exposure to PKC activators (PMA and PDBu) occluded the PGE_2 effect on TTX-R currents. These data suggest that

PKC activity is necessary to allow PKA-mediated modulation of TTX-R currents by PGE_2. This scenario is reminiscent of the permissive role of PKC mediated phosphorylation on PKA induced changes in brain neuron TTX-S currents except that the phopshorylation potentiates TTX-R currents but decreases TTX-S currents.

Phosphorylation of cloned $Na_v1.8$ channels

The likely sites for PKA modulation of this TTX-R Na^+ channel have been investigated by site-directed mutagenesis (Fitzgerald et al 1999). The $Na_v1.8$ subunit, like other types of Na^+ channel α subunits noted above, has five consensus PKA phosphorylation sites on the I–II intracellular linker loop. *In vitro* PKA-induced phosphorylation and tryptic peptide mapping of the channel confirmed that these five serine residues were the major PKA substrates in this region of the channel. Exposure of heterologously expressed wild-type $Na_v1.8$ channels to either forskolin or 8-bromo-cAMP resulted in a hyperpolarizing shift in the current–voltage relationship for activation and an increase in current amplitude evoked by test depolarizations to +5 to +20 mV. In contrast, mutant $Na_v1.8$ channels with the five serine residues replaced with alanine, were not significantly affected by either forskolin or 8-bromo-cAMP. These results illustrate the similarity of the response of native and cloned TTX-R currents to agents that raise cAMP levels and indicate that the effect is due to a PKA-induced phosphorylation of one or more of the serine residues on the I–II loop.

TTX-R modulation by other hyperalgesic agents

Two other mediators that are produced in inflammatory conditions, adenosine and serotonin, can also modulate TTX-R currents in DRG neurons (Gold et al 1996). In both cases, the mediators increased the size of the TTX-R current and shifted the current–voltage relationship for activation in the hyperpolarizing direction. Pharmacological characterization indicated that the serotonin effect was mediated by 5-HT_4 receptors (Cardenas et al 1997), which are typically linked to activation of adenylate cyclase.

Effects of *Herpes simplex* virus (HSV) on DRG neuron ion channels

HSV induces a selective loss of sodium currents in DRG neurons

Herpes simplex virus type 1 (HSV-1) is a common neurotropic virus that, *in vivo*, forms a latent infection in primary afferent neurons. HSV infection is associated with abnormal sensations around the site of initial infection including tingling, parasthesia, loss of touch and pain sensations (Andoh et al 1995) which have been

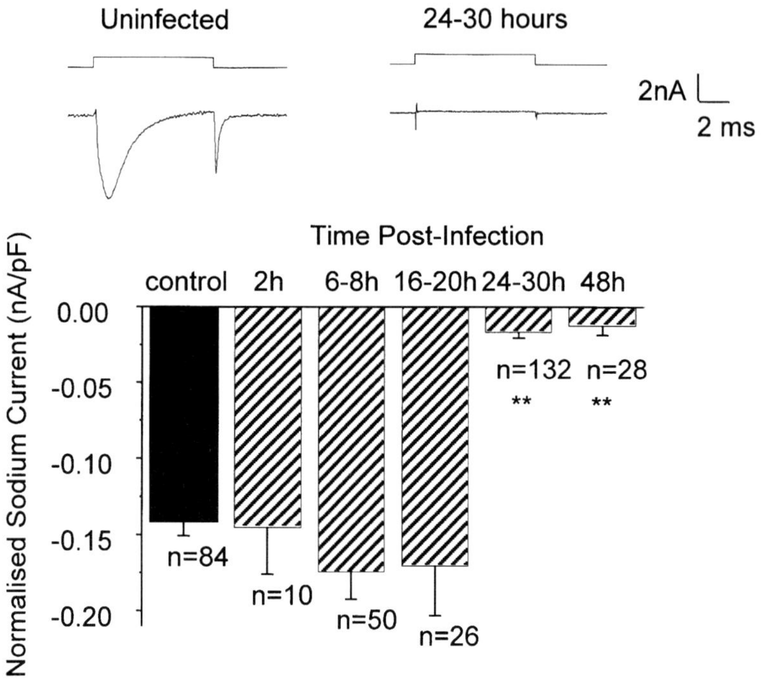

FIG. 4. (Top) Na^+ currents evoked by voltage steps from −80 mV to −10 mV in control (left) and HSV-infected DRG neurons (right). (Bottom) Normalized (nA/pF) Na^+ current amplitudes with above voltage step in control (filled bar) and at various times after HSV (5 plaque forming units/cell) infection show dramatic loss at 24 h.

attributed to alterations in neuronal excitability. HSV infection of DRG neurons *in vitro* abolishes the excitability of most neurons, and the neurons that remain excitable show action potentials that are smaller in amplitude and longer in duration than normal (Fukuda & Kurata 1981, Mayer et al 1986). Studies of the effects of HSV-1 infection on voltage gated Na^+ currents *in vitro* (Storey et al 1996) have shown that 24 h after HSV-1 infection a voltage step from −80 mV to 10 mV evoked a current in only 28% of neurons and that the amplitude of the current in these neurons was smaller than in uninfected neurons (HSV: -0.017 ± 0.004 nA/pF; control: -0.14 ± 0.009 nA/pF; $n=84$). Although no change in the Na^+ current amplitude was noted over the first 2–20 h after infection, the Na^+ current amplitude reduced rapidly and the reduction was maximal after 24 h of infection (Fig. 4). This loss of current was maintained until at least 48 h post-infection. No voltage-gated Na^+ currents were unmasked with voltage steps to a wide range of potentials from even more negative holding potentials indicating that the absence of currents was not due to any change in the voltage sensitivity of activation or inactivation. The

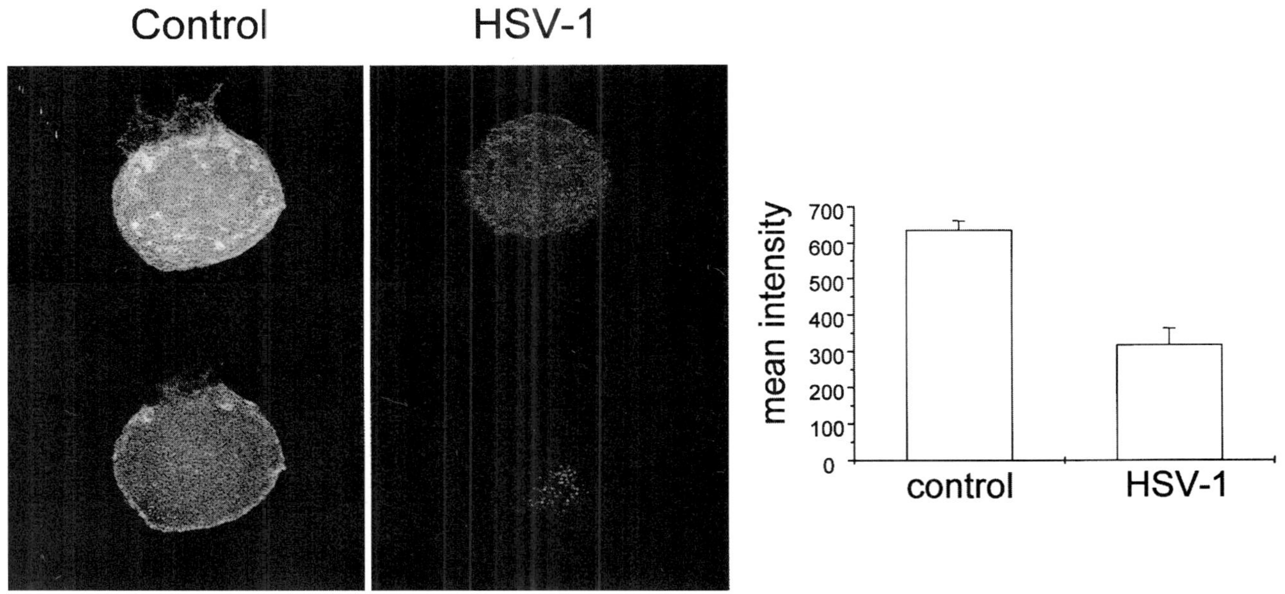

FIG. 5. (Left) Pan Na^+ channel antibody staining of control and HSV-infected (24 h) DRG neurons (top, whole cell staining; bottom, cross section image). (Right) Quantified immunofluorescence of Na^+ channel antibody staining.

HSV-1 induced loss of Na^+ currents was similar for both TTX-S and TTX-R currents. The effects of HSV-1 infection on voltage-gated ion channel function appeared to be restricted to Na^+ channels as Ca^{2+} currents and outwardly rectifying K^+ currents were not significantly changed after HSV-1 infection *in vitro*.

HSV-induced loss of Na^+ currents involves membrane internalization

Exposure of DRG neurons to inhibitors of membrane protein internalization (either bafilomycin A or chloroquine) prevented the loss of Na^+ conductance normally observed after HSV-1 infection (Storey et al 1998). Furthermore, confocal microscopy of DRG neurons stained with a pan Na^+ channel antibody showed that HSV-1 infection resulted in a marked loss of plasma membrane staining and an overall reduction in immunofluorescence throughout the neurons (Fig. 5). These findings are consistent with the hypothesis that HSV-1 induces Na^+ channel internalization. A possible mechanism of channel loss is suggested by experiments on veratridine-induced loss of surface Na^+ channels in neonatal (but not adult) rats CNS neurons where a rapid internalization of Na^+ channels ($T_{1/2}$ ~15 min) occurs after a rise in intracellular Na^+ concentration (Dargent & Couraud 1990, Dargent et al 1994). HSV-1 induced loss of Na^+ currents in adult rat DRG neurons showed a similar dependence on extracellular Na^+ and was inhibited when extracellular Na^+ was substituted with choline, which raises the possibility that a rise in internal Na^+ can mediate the HSV-1-induced internalization.

References

Andoh T, Shiraki K, Kurokawa M, Kuraishi Y 1995 Paresthesia induced by cutaneous infection with herpes simplex virus in rats. Neurosci Lett 190:101–104

Cantrell AR, Smith RD, Goldin AL, Scheuer T, Catterall WA 1997 Dopaminergic modulation of sodium current in hippocampal neurons via cAMP-dependent phosphorylation of specific sites in the sodium channel α subunit. J Neurosci 17:7330–7338

Cantrell AR, Yu F, Sharp EM et al 2000 Molecular mechanisms underlying convergent regulation of brain Na^+ channels by protein kinase A and protein kinase C. Soc Neurosc Abstr 26:1908

Cardenas CG, Del Mar LP, Cooper BY, Scroggs RS 1997 $5HT_4$ receptors couple positively to tetrodotoxin-insensitive sodium channels in a subpopulation of capsaicin-sensitive rat sensory neurons. J Neurosci 17:7181–7189

Chen T-C, Law B, Kondratyuk T, Rossie S 1995 Identification of soluble protein phosphatases that dephosphorylate voltage-sensitive sodium channels in rat brain. J Biol Chem 270:7750–7756

Costa MR, Catterall WA 1984a Cyclic AMP-dependent phosphorylation of the alpha subunit of the sodium channel in synaptic nerve ending particles. J Biol Chem 259:8210–8218

Costa MR, Catterall WA 1984b Phosphorylation of the alpha subunit of the sodium channel by protein kinase C. Cell Mol Neurobiol 4:291–297

Dargent B, Couraud F 1990 Down-regulation of voltage-dependent sodium channels initiated by sodium influx in developing neurons. Proc Natl Acad Sci USA 87:5907–5911
Dargent B, Paillart C, Carlier E, Alcaraz G, Martin-Eauclaire MF, Couraud F 1994 Sodium channel internalization in developing neurons. Neuron 13:683–690
England S, Bevan S, Docherty RJ 1996 PGE2 modulates the tetrodotoxin-resistant sodium current in neonatal rat dorsal root ganglion neurones via the cyclic AMP-protein kinase A cascade. J Physiol 495:429–440
Fitzgerald EM, Okuse K, Wood JD, Dolphin AC, Moss SJ 1999 cAMP-dependent phosphorylation of the tetrodotoxin-resistant voltage-dependent sodium channel SNS. J Physiol 516:433–446
Fukuda J, Kurata T 1981 Loss of membrane excitability after herpes simplex virus infection in tissue-cultured nerve cells from adult mammals. Brain Res 211:235–241
Gold MS, Reichling DB, Shuster MJ, Levine JD 1996 Hyperalgesic agents increase a tetrodotoxin-resistant Na^+ current in nociceptors. Proc Natl Acad Sci USA 93:1108–1112
Gold MS, Levine JD, Correa AM 1998 Modulation of TTX-R I_{Na} by PKC and PKA and their role in PGE_2-induced sensitization of rat sensory neurons in vitro. J Neurosci 18:10345–10355
Li M, West JW, Lai Y, Scheuer T, Catterall WA 1992 Functional modulation of brain sodium channels by cAMP-dependent phosphorylation. Neuron 8:1151–1159
Li M, West JW, Numann R, Murphy BJ, Scheuer T, Catterall WA 1993 Convergent regulation of sodium channels by protein kinase C and cAMP-dependent protein kinase. Science 261:1439–1442
Mayer ML, James MH, Russell RJ, Kelly JS, Pasternak CA 1986 Changes in excitability induced by herpes simplex viruses in rat dorsal root ganglion neurons. J Neurosci 6:391–402
Murphy BJ, Rossie S, De Jongh KS, Catterall WA 1993 Identification of the sites of selective phosphorylation and dephosphorylation of the rat brain Na^+ channel alpha subunit by cAMP-dependent protein kinase and phosphoprotein phosphatases. J Biol Chem 268:27355–27362
Numann R, Catterall WA, Scheuer T 1991 Functional modulation of brain sodium channels by protein kinase C phosphorylation. Science 254:115–118
Smith RD, Goldin AL 1996 Phosphorylation of brain sodium channels in the I–II linker modulates channel function in *Xenopus* oocytes. J Neurosci 16:1965–1974
Storey NM, Latchman DS, Bevan S 1996 TTX-sensitive (TTXs) and TTX-resistant (TTXr) voltage-gated sodium currents in adult rat dorsal root ganglion neurones are inhibited by a herpes simplex virus replication product in vitro. J Physiol 495:72P
Storey NM, Latchman DS, Bevan S 1998 Internalisation of voltage gated sodium channels in adult rat dorsal root ganglion neurones infected with Herpes simplex virus in vitro. Soc Neurosci Abstr 24:380
West JW, Numann R, Murphy BJ, Scheuer T, Catterall WA 1992 Phosphorylation of a conserved protein kinase C site is required for modulation of Na^+ currents in transfected Chinese hamster ovary cells. Biophys J 62:31–33

DISCUSSION

Gold: I have a question about the K^+ currents. In your paper describing modulation of TTX-R Na^+ currents in DRG neurons, you also described the inhibition of K^+ currents (England et al 1996). Grant Nicol and Michael Vasko reported a similar inflammatory mediator-induced inhibition of K^+ currents, but the time course they reported was relatively slow (Nicol et al 1997). Furthermore, the inhibition of K^+ currents that you described was reversed following wash of PGE_2. Because PGE_2-induced changes in excitability occur relatively quickly and are not very reversible following wash of PGE_2, how do these observations fit with inhibition of K^+ current as an underlying mechanism of sensitization?

Bevan: I can speculate. As far as the underlying mechanism for a wash, they weren't very rapid reversible effects. If phosphorylation of the K^+ channels is involved in this, it will obviously depend on the dynamics of the dephosphorylation, and the balance between the phosphorylated and dephosphorylated states.

Catterall: It is striking how different and opposite the effects of PKA and PKC phosphorylation are on these Na^+ channels from DRG neurons, compared with the type 2A channels that we have studied that are expressed in the brain. In those we have studied, the same signalling pathways (the PKA and PKC pathways) reduce channel activity, but the PKA and PKC act synergistically to do so, just as they do in your experiments to increase channel activity with the SNS channels. It is as if the sign of the regulation is different, but the underlying mechanisms are the same. In the type 2 channel, it is the phosphorylation sites in the loop between domains 1 and 2 that are important, as both Al Goldin's lab and ours have shown. It is as if these two channels are set up to respond to the signalling pathways differently, but may do so by a common mechanism.

Bevan: It will be fascinating to do some of the experiments that you have done involving site-directed mutagenesis to pin down the residues involved in the SNS channel.

Catterall: That story has become even more complicated in more recent experiments. We have found that both membrane potential and PKC influence the functional effects caused by phosphorylating individual PKA sites.

Spruston: Have you compared whether the effect is the same or different in the TTX-S Na^+ channels with these cells? It could also be a cell-specific effect as opposed to a channel-specific effect.

Bevan: We studied neonatal rat DRG neurons, and most of the cells we studied seemed to respond to prostanoids. This is not the case in the adult cells, where about 50% of them don't respond. If you take cells that only have TTX-S currents, then there doesn't seem to be much modulation at all by prostanoids. We didn't look at cAMP analogues in those experiments. During the separation of TTX-R and TTX-S, it looks as though there may be an effect by prostanoids on the TTX-S currents in those cells that showed a TTX-R modulation. We didn't study this in detail; it seemed to be much more subtle than the effect on the TTX-R current.

Gold: That is one of the nice things about the DRG neurons: you can study both currents and use one as a control for the other. It wasn't consistent that we would see an inhibition of the TTX-S current, but we did see an inhibition as a general rule. This was consistent with the brain-type channels. While PGE_2 would increase TTX-R it would decrease the TTX-S. It didn't seem to be cell specific.

Bevan: It is a rather nice system where there are opposite effects in the same cell by activating the same pathways.

Goldin: Are the effects reversible, and with a similar time course?

Bevan: In our experiments we didn't study the cells for a long time. In some of them we saw the currents begin to decrease. We were always worried about long-term rundown in the currents. They are reasonably robust but after an hour I wouldn't be confident we weren't getting effects completely unrelated to previous drug treatment.

Bean: How do you choose the cells in doing these experiments? You said these were from neonates.

Bevan: Ours were from neonates and Michael Gold's were from adults. For ours, we didn't use any special criteria to select the cells.

Bean: One hears so much about the heterogeneity of sensory neurons. Is this an issue with the adult cells?

Gold: They are heterogeneous. Early on we had broken them into categories on the basis of cell size, responsiveness to capsaicin and IB4 binding. In general it doesn't seem to fall out in terms of which population will be modulated and which won't by any of those categories. Our interpretation of that is that it had to do more with how the receptor for the prostaglandin is distributed. The receptor didn't seem to be distributed with any a priori category we used. With respect to cell size, my feeling is that it is much less predictive than some people have been proposing. There is a lot of weight put on the functional significance of cell size. There is a loose correlation that smaller cell bodies give rise to more slowly conducting axons. I think this is reflective of target innervation. This correlation holds up fairly well with cutaneous afferents, but in the visceral afferents that we have recently been studying, the correlation with cell body size seems to fall down. It should be noted that Reese Scroggs has implemented a system for categorizing DRG neurons and he has reported that specific groups of cells respond to particular inflammatory mediators (Cardenas et al 1997).

Waxman: Bruce Bean is raising a very important point. Cutaneous afferents express large TTX-R currents, whereas muscle afferents show lower levels. Then we have the TTX-S fast current which is probably the composite of the products of several different TTX-S channels. The reflex solution is to put them into heterologous expression systems and Lori Isom addressed very clearly the problems we run into there. We need to triangulate between studying native systems and the various expression systems.

Catterall: However, the PKC effect is exactly the same in the transfected system as it is in hippocampal neurons.

Goldin: You said you saw variability in the decrease of the TTX-S currents. This would make sense, because 1.1 is decreased by PKA, whereas 1.6 does not have the critical second PKA site, and as predicted from this absence it does not show a decrease with PKA. If there are variable proportions of 1 and 6, you would see variable amounts of decreased current.

Strichartz: I have some concern about the steady state levels of phosphorylation being very dependent on the activity of phosphatases. I wonder whether these experiments should be repeated with perforated patch or cell-attached patches, because of the possibility of a much larger or smaller effect as a result of the inevitable perfusion that occurs. Does anyone know anything about the effects of the intracellular media that we choose? They are pretty much the same.

Gold: I think that is a good point. Using a cell-attached patch we have seen larger modulation. The other issue that also adds to this is the complication to the story with PKA and PKC when you move back to *in vivo* models. For instance, people who have been trying to inhibit PGE_2-induced hyperalgaesia with a PKC antagonist have been unable to do so. This is associated with what was shown in terms of the *in vitro* data, although there are inflammatory mediators that do act through a PKC pathway. Our subsequent data would suggest that this reflects the level of resting Ca^{2+} in the cells, which changes depending on how you prepare your cells. If we do Ca^{2+} measurements on cells prepared with one sort of enzyme versus another and look at neurons in a window of 4–6 h versus 18–20 h, the resting Ca^{2+} levels change. Consequently, my guess is that resting PKC activity and phosphorylation is going to change. This will impact what is seen in terms of subsequent modulation of Na^+ currents.

Strichartz: What is the highest resting Ca^{2+} level that you see?

Gold: If I recall correctly, about 200 nM.

Bean: It is extremely difficult to do Na^+ channel voltage clamping with perforated patch recordings. It is hard to get the series resistance down low enough to get a good clamp.

Raman: In whole cell recordings from Purkinje cells over time, the inactivation curve tends to shift to more negative voltages.

Strichartz: That is a universal phenomenon, by the way.

Horn: Not so much in two microelectrodes in oocytes.

Strichartz: It seems to occur when you start mucking up the cytoplasm.

Horn: In perforated patches it tends to be relatively stable.

Strichartz: Interestingly enough, in old experiments that Jim Fox did in node of Ranvier, if he set the holding potential very negative (to about −110 mV) he found that the mid-point for inactivation shifted much more slowly over the same period than if the membrane were held at −80 mV (Fox 1976). So it seems that this 'drift' does depend on the state of the Na^+ channel. If you can remove slow inactivation, you don't see this drift in the fast inactivation parameter.

Horn: In our experiments we usually use holding potentials of -140 or -150 mV, and we still see those shifts. There isn't much slow inactivation at those holding potentials.

Wood: I was very interested in the HSV data. I was rather surprised at these dramatic effects, because one normally associates HSV with a latent infection, with very low level viral protein synthesis. Does this *in vitro* model reflect that form of latency, or is it a high multiplicity of infection model?

Bevan: It is a high multiplicity of infection. We have 5 pfu per DRG.

Wood: Did the cells survive long-term?

Bevan: No. We were interested in trying to induce latency and then looking at recovery from latency, but *in vitro* there are no reliable methods of doing that.

Noebels: Are these replication-defective HSVs being used as vectors for gene transfer?

Bevan: Nina Storey studied a large number of vectors that are potentially going to be used therapeutically, and none of them have effects on any of the currents.

Cummins: We tried the amplicon system and it seemed to have major effects on the Na^+ currents in both DRG and cortical neurons

Bevan: It depends on which vectors you are using. The ones we looked at were those developed by David Latchman and intended ultimately for therapeutic use.

Strichartz: We were doing some experiments on neuroblastoma cells to try to understand local anaesthetic-induced neurotoxity, which was a clinical problem a while ago. Anaesthesiologists use obscenely high concentrations of local anaesthetics because the efficiency of transport into the peripheral nerves is so low; 200 mM is not unusual. We applied 200 mM lidocaine and saw that although some of the cells died, many didn't. When lidocaine was removed after an hour, all the cells that didn't have crenulated membranes and that we could get a whole seal on, had almost no current in them. However, over the next 10 h the current grew back to normal (of course we were not clamping any one cell for 10 h). We thought that this might be due to the independently observed increase in intracellular Ca^{2+} that is produced by local anaesthetics, and which I think you have also seen in DRG. I wondered whether there might not be Ca^{2+} going through your Na^+ channels and producing those changes, and it is not Na^+ per se, but an elevation of intracellular Ca^{2+}.

Bevan: We can't rule that out, because we haven't checked the Ca^{2+} levels. One can get a similar down-regulation of the Na^+ current in these cells by short-term treatment with veratradine.

Meisler: I wanted to ask about the brefeldin experiment you used as a control. Does that prevent turnover or delivery of the endogenous proteins as well? If so, does that say anything about the turnover of channels?

Bevan: It will block all proteins being exported.

Meisler: It had no effect on channel activities in 24 h.

Bevan: That would fit in with cycloheximide experiments, where over that period we see no appreciable loss of Na^+ currents. Whether preexisting channels can be cycling in and out is another question.

Catterall: Francois Couraud's lab in Marseilles has done similar experiments on cultured brain neurons from early embryos, and finds a similar Na^+ dependent internalization and down-regulation of Na^+ channels. However, this is not observed in more mature cultures of neurons. Do you see the same correlation between the age of the neurons and the ability to down-regulate Na^+ channels?

Bevan: All the HSV experiments were done on adult neurons. On the basis of their results we were expecting not to see an effect of veratradine in the adult animals, but in fact we clearly do see this. It occurs within about 5 h, which is very similar to the time course they see for their cortical neurons from young rats. At one stage they speculated that the stability changed: as the animals aged, the veratradine effect was lost. They speculated this was due to an increase in the synthesis of β subunits. I don't know whether there is any further evidence for or against that hypothesis.

Catterall: There's a temporal correlation between expression of β subunits and loss of down-regulation. Over the same time frame the Na^+ channels also change from being mostly type 3 to mostly type 1 or 2. Couraud and colleagues found that it is Na^+ and not Ca^{2+} that is the mediator. Other than this, I don't think the mechanism is further developed.

References

Cardenas CG, Del Mar LP, Cooper BY, Scroggs RS 1997 5HT4 receptors couple positively to tetrodotoxin-insensitive sodium channels in a subpopulation of capsaicin-sensitive rat sensory neurons. J Neurosci 17:7181–7189

England S, Bevan S, Docherty RJ 1996 PGE2 modulates the tetrodotoxin-resistant sodium current in neonatal rat dorsal root ganglion neurons via the cyclic AMP–protein kinase A cascade. J Physiol 495:429–440

Fox JM 1976 Ultra-slow inactivation of the ionic currents through the membrane of myelinated nerve. Biochim Biophys Acta 426:232–244

Nicol GD, Vasko MR, Evans AR 1997 Prostaglandins suppress an outward potassium current in embryonic rat sensory neurons. J Neurophysiol 77:167–176

Sodium channels in primary sensory neurons: relationship to pain states

John N. Wood, Armen N. Akopian, Mark Baker, Yanning Ding, Fleur Geoghegan, Mohammed Nassar, Misbah Malik-Hall, Kenji Okuse, Louisa Poon, Samantha Ravenall, Madhu Sukumaran and Veronika Souslova

Department of Biology, Medawar Building, University College London, Gower Street, London WC1E 6BT, UK

Abstract: Electrophysiological studies of dorsal root ganglion (DRG) neurons, and the results of PCR, Northern blot and *in situ* hybridization analyses have demonstrated the molecular diversity of Na^+ channels that operate in sensory neurons. Several subtypes of α-subunit have been detected in DRG neurons and transcripts encoding all three β-subunits are also present. Interestingly, one α subunit, $Na_v1.8$, is selectively expressed in C-fibre and Aδ fibre associated sensory neurons that are predominantly involved in damage sensing. Another channel, $Na_v1.3$, is selectively up regulated in a variety of models of neuropathic pain. In this review we focus on Na^+ channels that are selectively expressed in DRG neurons as potential analgesic drug targets. In the absence of subtype specific inhibitors, the production of null mutant mice provides useful information on the specialized functions of particular Na^+ channels. A refinement of this approach is to delete Na^+ channel genes flanked by lox-P sites in the sensory ganglia of adult animals, using viruses to deliver the bacteriophage Cre recombinase enzyme.

2002 Sodium channels and neuronal hyperexcitability. Wiley, Chichester (Novartis Foundation Symposium 241) p 159–172

Damage-sensing neurons express Na^+ channels with unusual pharmacology. The application of molecular genetics has allowed the identification of the individual subunits that underlie voltage-gated Na^+ channel activity. Interestingly, there do seem to be Na^+ channel isoforms that are specifically associated with damage-sensing neuron function, making these channels attractive targets for pharmacological intervention to produce analgesia (Clare et al 2000). The regulation of Na^+ channel expression by accessory subunits (Isom 2002, this volume), and their activity by phosphorylation (Bevan 2002, this volume) are topics of considerable relevance in terms of developing analgesic drugs.

The functional heterogeneity of Na^+ channels in sensory neurons was detected using the channel blocking activity of the puffer fish poison tetrodotoxin (TTX)

(Matsuda et al 1978). Na^+ channels expressed in small-diameter dorsal root ganglion (DRG) neurons are easily separated into TTX-sensitive (TTX-S) and TTX-resistant (TTX-R) components. A number of groups have provided functional evidence for the diversity of TTX-S currents in DRG neurons, related to cell diameter (Caffrey et al 1992, Elliott & Elliot 1993, Rush et al 1998). Ogata & Tatebayashi (1992) showed that in the adult rat, large diameter neurons expressed TTX-S Na^+ currents, whilst the smallest diameter neurons expressed predominantly TTX-R Na^+ currents. About a third of the total number of cells expressed both types of current, and their mean diameter was intermediate between the large and small populations. Studies utilizing mRNA *in situ* hybridization, Northern blots and RT-PCR also imply that several α-subunits may be involved in generating TTX-S currents in sensory neurons (reviewed in Baker & Wood 2001). High-threshold TTX-R currents seem to be generated by the $Na_v1.8$ channel as they have not been detected in murine $Na_v1.8$ knockout neurons (Akopian et al 1999a). The current attributable to the NaN or SNS-2 channel ($Na_v1.9$) can be conveniently examined in $Na_v1.8$ null mutant DRG neurons (Cummins et al 1999). The voltage dependence of $Na_v1.8$ currents suggests that this channel would be largely or entirely inactivated at the normal resting potential of small diameter neurons. The expression of voltage-gated Na^+ channel α-subunits in sensory neurons is summarized in Table 1.

Recent functional evidence suggests that TTX-R currents are involved in the transduction of noxious stimuli in the cornea and the intracranial dura, whilst

TABLE 1 Voltage-gated Na^+ channel α subunits of sensory neurons

Channel	*Previous name*	*Gene symbol*	*Chromosome (human)*	*Pharmacology*	*Abundance in adult DRG*
$Na_v1.1$	Type I	*SCN1A*	2q24	TTX-S	Present
$Na_v1.2$	Type II	*SCN2A*	2q23-24	TTX-S	Present
$Na_v1.3$	Type III	*SCN3A*	2q24	TTX-S	Up-regulated in axotomy
$Na_v1.4$	SkM	*SCN4A*	17q23-25	TTX-S	Absent
$Na_v1.5$	Cardiac	*SCN5A*	3p21	TTX-R	Absent
$Na_v1.6$	NaCh6	*SCN8A*	12q13	TTX-S	Abundant
$Na_v1.7$	PN1	*SCN9A*	2q24	TTX-S	Abundant
$Na_v1.8$	SNS/PN3	*SCN10A*	3p21-24	TTX-R	Abundant
$Na_v1.9$	NaN	*SCN11A*	3p21-24	TTX-R	Abundant
Na_x	NaG	*SCN6A/ SCN7A*	2q21-23	?	Present

TTX-S channels are critical for action potential transmission in many of the innervating unmyelinated fibres (Brock et al 1998, Strassman & Raymond 1999). In experiments on the mechanical sensitivity of the dural membrane, the application of 1 μM TTX to the outer face of the membrane completely suppressed responses to mechanical stimulation in larger Aδ fibres, whereas 50% of small Aδ and more than 85% of the C-fibres recorded were unaffected, indicating that TTX-R channels are an important element of transduction in the smallest endings. In experiments on mechanically activated corneal units, the additional step of blocking Ca^{2+} channels using Cd^{2+} suggests TTX-R Na^+ channels are involved in transduction and/or action potential generation at the endings. In human sural nerve, conduction in some C-fibres can continue in the presence of 10 μM TTX, although action potential amplitude is reduced between 20–80% (Grosskreutz et al 1996). In some axons TTX-R Na^+ currents are thus sufficient to propagate action potentials, but TTX-S channels must be essential for transmission in most C-fibres.

Regulation of expression of Na^+ channels

β1 and β2 subunits that associate with α subunits were first isolated by biochemical purification procedures (Hartshorne et al 1982). Molecular cloning has recently identified a splice variant, β1a, and a novel β1-like member of the family, βIII, which shows an overlapping or complementary pattern of expression to the β1 subunit (Catterall 2000, Isom 2000, Kazen-Gillespie et al 2000, Malhotra et al 2000, Morgan et al 2000). Early studies focused on the role of these subunits in enhancing levels of channel expression and altered kinetic properties, recent evidence suggest that β subunits may play an important role as cell adhesion molecules in determining the spatial pattern of expression of functional α subunits (Isom 2000). As well as the increasing complexity of accessory subunits linked to cytoskeletal and extracellular elements, there is evidence for the existence of multiprotein complexes involving tyrosine kinases. Voltage-gated Na^+ channels in brain neurons were found to associate with receptor protein tyrosine phosphatase β (RPTP-β) (Ratcliffe et al 2000). Na^+ channels were found to be tyrosine phosphorylated and dephosphorylation slowed $Na_v1.1$ inactivation, positively shifting voltage dependence, and increasing whole-cell Na^+ current. These aspects of channel regulation are of considerable interest in terms of the development of analgesics, because the development of sub-type selective blockers has proved elusive. Down-regulation of expression or modulation of function may provide an alternative route to effecting analgesia. The TTX-R channel $Na_v1.8$ is poorly expressed in heterologous expression systems, suggesting a requirement for accessory proteins to promote efficient expression of this channel. Disrupting interactions with permissive factors required for high-level channel expression maybe a useful approach to inhibiting channel function.

Altered Na^+ channel activity in pain states

Strong evidence suggests that altered Na^+ channel activity plays an important role in both inflammatory and neuropathic pain. Altered patterns of Na^+ channel transcripts, as well as post-translational modifications of α subunits have been observed (Black et al 1999, Waxman et al 1999, Baker & Wood 2001). Inflammatory pain is associated with lowered thresholds of activation of nociceptors in the periphery and altered TTX-R functional activity has been proposed to underlie some elements of this phenomenon (Khasar et al 1998, Gold 1999). A variety of hyperalgesic mediators that alter pain thresholds (i.e. prostaglandin E_2, serotonin) increase TTX-R currents in dissociated sensory neurons, and shift the activation voltage-dependence to more negative potentials (Gold et al 1996, England et al 1996) Good evidence that protein kinase A activation underlies these events has been obtained (Fitzgerald et al 1999).

These phenomena can be replicated in heterologous expression systems, where the properties of the $Na_v1.8$ channel are modulated by increasing cAMP levels, suggesting that the $Na_v1.8$ channel underlies the changes in TTX-R activity seen in sensory neurons. A major role for the TTX-R Na^+ current in conferring excitability upon nociceptive nerve endings could explain why some nerve endings are silent except when the surrounding tissues are inflamed.

Apart from small inflammatory mediators such as serotonin and prostaglandins, there is also evidence that nerve growth factor (NGF) is a key regulator of nociceptive thresholds. NGF causes dramatic decreases in thermal thresholds of pain perception in animal models and NGF levels rise in damaged tissue (Lewin et al 1993). NGF acts through a high affinity receptor TrkA to induce rapid changes in gene expression. There is evidence for transcriptional regulation of both α and $\beta1$ subunit expression by NGF in embryonic DRG neurons (Zur et al 1995). With the addition of NGF, DRG neurons express moderate levels of $Na_v1.1$ and $Na_v1.3$ subunits, and high levels of $Na_v1.2$ transcript. Na^+ channel $\beta1$ mRNA was also up-regulated by NGF to a high level of expression. Interestingly, the TTX-R channel $Na_v1.8$, whilst apparently requiring tonic NGF treatment for expression, is not dramatically up-regulated by high levels of NGF. However, an unusual, apparently non-functional splice variant is induced by NGF treatment, and appears to arise by the novel mechanism of *trans*-splicing, where three exons are repeated in the splice variant (Akopian et al 1999b). Over-expression of NGF in transgenic mice results in large increases in $Na_v1.2$ and $\beta2$ transcripts, whilst mRNA levels for $Na_v1.1$, Na_x, $Na_v1.6$, $Na_v1.8$, $Na_v1.9$ and $\beta1$ were also greater (Fjell et al 1999).

Neuropathic pain that results from direct damage to peripheral nerves is the most problematic condition in terms of analgesic therapy, and seems to depend in part upon dysregulation of Na^+ channel expression. The pain evoked by these conditions is associated with ectopic TTX-S action potential propagation initiated at the site of nerve injury (Lyu et al 2000). After chronic constriction injury Na^+ channel activity in acutely dissociated sensory neurons showed reduced levels of expression of TTX-R Na^+ channels and the functional properties of the TTX-R channels that are expressed were altered (Dibb-Hajj et al 1999, Kral et al 1999). The current–voltage relationship shifted to give a change in both activation and steady-state inactivation properties of the total Na^+ current to more negative potentials without a significant change in the density of total Na^+ current. Such changes may contribute to neuronal excitability associated with this form of nerve damage. When peripheral nerves are severed or damaged, there is a loss of expression of $Na_v1.8$ and $Na_v1.9$ transcripts, but an increase in the type $Na_v1.3$ Na^+ channel (Waxman et al 1994, Black et al 1999). Ectopic application of glial-derived neurotrophic factor (GDNF) can reverse these changes. GDNF acts through an unusual complex of transforming growth factor (TGF)β receptor like subunits and the c-Ret tyrosine kinase, which is expressed on subsets of both small and large diameter sensory neurons. GDNF not only reverses the changes in Na^+ channel expression patterns characteristic of damaged sensory neurons, but is also able to reverse neuropathic pain behaviour in animal models. This suggests that the $Na_v1.3$ channel, normally expressed only in developing sensory neurons, may play an important role in the pathogenesis of neuropathic pain (Boucher et al 2000).

A direct approach to determine the functional significance of Na^+ channel isoforms and accessory subunits is to ablate the expression of channels in null mutant mice, and measure the behavioural and electrophysiological consequences. So far three Na^+ channel null mutant mice have been described. The $Na_v1.2$ null mutant dies shortly after birth, with massive brain-stem apoptosis leading to respiratory problems (Planells-Cases et al 2000). Na^+ channel currents recorded from cultured neurons of $Na_v1.2$ null mice are sharply attenuated. $Na_v1.2$ expression is thus redundant for embryonic development but essential for postnatal survival. An analysis of the significance of this channel in pain behaviour is thus precluded. The Na_x channel, for which there is no functional data demonstrating Na^+ channel activity, is viable and shows deficits in its regulation of salt intake, although its pain behaviour has not been analysed (Watanabe et al 2000) The deletion of the $Na_v1.8$ gene encoding a sensory neuron specific TTX-R channel does not effect the viability of null mutant mice. Studies of such null mutants demonstrate that all TTX-R activity found in sensory neurons apart from a persistent Na^+ current, probably corresponding to NaN is encoded by $Na_v1.8$ (Akopian et al 1999a, Cummins et al 1999). The loss of

$Na_v1.8$ seems to lead to a compensatory up-regulation of expression of the TTX-S channel PN1 ($Na_v1.7$), with lower thresholds of electrical excitation of C-fibres in null mutants. There are nonetheless major deficits in pain pathways in these null mutants, in particular in responses to noxious mechanical stimulation emphasizing the important role of $Na_v1.8$ in nociception (Akopian et al 1999a). Interestingly, low-dose systemic lidocaine dramatically enhances the analgesic phenotype of the mutant mouse at concentrations that do not affect wild-type mouse behaviour. As TTX-S currents are more sensitive to lidocaine than TTX-R currents (Roy & Narahashi 1992, Akopian et al 1999a), this effect suggests that the analgesic phenotype of the SNS null mutant would be much more dramatic but for the compensatory up-regulation of TTX-S channels in these animals. The generation of NaN knockouts and the analysis of the phenotype of the combined SNS/NaN knockouts will be particularly illuminating in assessing the significance of TTX-R currents for nociception and pain processing, and in proving the relationship between $Na_v1.9$ expression and the presence of a persistent TTX-R current.

Despite considerable efforts to develop sub-type specific Na^+ channel blockers, there are as yet no highly specific reagents available (Clare et al 2000). Given the global expression patterns within the CNS of many Na^+ channel isoforms, the generation of null mutant mice is likely to lead to phenotypes whose significance for pain processing in the periphery is hard to evaluate. A possible route to obtaining useful information on this topic is to generate mice in which Na^+ channel genes are only ablated within sensory ganglia. This approach is now feasible thanks to the work of Sauer and collaborators (Le & Sauer 2000), who have exploited the recombinase activity of a bacteriophage enzyme Cre, to delete DNA sequences that are flanked by lox-P sites recognized by this enzyme. Producing mice with lox-P-flanked functional Na^+ channel genes should thus allow the tissue specific ablation of these genes, depending upon the tissue in which functional Cre recombinase activity is expressed in transgenic mice. In order to ablate genes in sensory ganglia, it is necessary to produce mice in which functional Cre recombinase is driven by sensory neuron-specific promoters. In order to identify such useful regulatory elements, transgenic mice expressing reporter genes such as *lacZ* driven by various promoters must be analysed. A short cut to identifying such elements involves transfecting sensory neurons with reporter constructs and analysing the cell types in which the reporter is expressed. In order to equalize transfection efficiency between the various cell types present in the cell cultures derived from DRG, a biolistic gene gun can be used. This approach depends upon shooting microscopic gold particles coated with DNA into the culture dish, where all cell types receive equivalent amounts of intracellular DNA. By measuring reporter gene expression, and then typing the subsets of cells that express a particular gene, some information that may be relevant to the pattern of gene expression in whole animals can be obtained. Transgenic mice that express Cre recombinase in the desired subsets of cells can then be constructed. Using promoters derived from mapped regulatory sequences in peripherin, c-Ret (the GDNF receptor kinase) and $P2X_3$, a sensory neuron-specific ATP-gated cation channel, a range of Cre-expressing mice are being characterized. In order to analyse the effectiveness of expressed Cre in excising lox-P-flanked genes, a reporter mouse using the β-galactosidase expressing gene with a floxed (lox-P flanked) stop signal can be used. Where Cre removes the stop signal β-galactosidase activity can be analysed histochemically. Transgenic mice that express Cre only in DRG neurons should prove useful in excising floxed targets such as the $Na_v1.3$ Na^+ channel implicated in the development of ectopic action potential propagation after nerve damage.

One remaining problem with tissue-specific null mutants that exploit the Cre/lox-P system is that early excision at the time of birth of sensory neurons may allow developmental compensatory mechanisms to occur, again complicating phenotypic analysis, particularly in terms of behaviour. It would thus be ideal to be able to delete genes in adult animals, whose behaviour under control conditions has already been assessed. Recent advances in gene delivery suggest that this approach may be feasible.

Herpes viruses type 1 and 2 are known to establish long lasting latent infections in sensory neurons, and are suitable vectors to deliver gene fragments up to about 15 kb. Using HSV expressing a fusion protein comprising green fluorescent protein tagged to Cre recombinase (GFP-Cre) we, and others have been able to deliver functional Cre to a variable number of sensory neurons in adult ROSA-26 reporter mice (Rinaldi et al 1999, Marshall et al 2000). It is important to use a replication-defective HSV mutant for such studies, so that peripheral tissue damage is minimised. Although this approach shows promise, the mosaicism in gene ablation within sensory ganglia is likely to make behavioural analysis tedious, because very large numbers of animals will be necessary to form an unambiguous view of the effects of gene deletion. With respect to the number of sensory neurons that can be infected with gene delivery vectors, it may be more useful to use adenovirus, or adenovirus-associated vectors, which show a broader level of cell tropism. Adenovirus, which can deliver up to 7 kb of extraneous DNA, expresses Cre recombinase activity more effectively, but infects both neuronal and non-neuronal cells. Such vectors have been constructed. If high-level gene delivery can be affected by this route, then the complementary approach of delivering antisense constructs to specifically down-regulate expression of particular transcripts may prove attractive.

In summary, the application of molecular genetic techniques has revolutionized our understanding of Na^+ channel structure and function. With recent developments in high expression cell-specific gene delivery systems, inducible knockouts, and the complete sequencing of the mouse and human genomes, it should be possible to answer many questions about the specialized role of Na^+ channel isoforms in pain processing in the near future.

Acknowledgements

We thank the Wellcome Trust and the MRC for support. We are very grateful to our collaborators for their help.

References

Akopian AN, Souslova V, England S et al 1999a The tetrodotoxin-resistant sodium channel SNS has a specialized function in pain pathways. Nat Neurosci 2:541–548

Akopian AN, Okuse K, Souslova V, England S, Ogata N Wood JN 1999b Trans-splicing of a voltage-gated sodium channel is regulated by Nerve Growth Factor. FEBS Lett 445:177–182

Bevan S 2002 Modulation of sodium channels in primary afferent neurons. In: Sodium Channels and neuronal hyperexcitability. Wiley, Chichester (Novartis Found Symp 241) p 144–158

Baker MD, Wood JN 2001 Involvement of Na channels in pain pathways. Trends in Pharmacology 22:27–31

Black JA, Cummins TR, Plumpton C et al 1999 Upregulation of a silent sodium channel after peripheral, but not central, nerve injury in DRG neurons. J Neurophysiol 82:2776–2785

Brock JA, McLachlan EM, Belmonte C 1998 Tetrodotoxin-resistant impulses in single nociceptor nerve terminals in guinea-pig cornea. J Physiol 512:211–217

Boucher TJ, Okuse K, Bennett DL, Munson JB, Wood JN, McMahon SB 2000 Potent Analgesic Effects of GDNF in Neuropathic Pain States. Science 290:124–127

Catterall WA 2000 From ionic currents to molecular mechanisms: the structure and function of voltage-gated sodium channels. Neuron 26:13–25

Caffrey JM, Eng DL, Black JA, Waxman SG, Kocsis JD 1992 Three types of sodium channels in adult rat dorsal root ganglion neurons. Brain Res 592:283–297

Clare JF, Tate SN, Nobbs M, Romanos MA 2000 Voltage-gated sodium channels as therapeutic targets. Drug Discovery Today 5:506–520

Cummins TR, Dib-Hajj SD, Black JA, Akopian AN, Wood JN, Waxman SG 1999 A novel persistent tetrodotoxin-resistant sodium current in SNS-null and wild-type small primary sensory neurons. J Neurosci 19:RC43

Dib-Hajj SD, Fjell J, Cummins TR et al 1999 Plasticity of sodium channel expression in DRG neurons in the chronic constriction injury model of neuropathic pain. Pain 83:591–600

Elliott AA, Elliott JR 1993 Characterization of TTX-sensitive and TTX-resistant sodium currents in small cells from adult rat dorsal root ganglia. J Physiol 463:39–56

England S, Bevan S, Docherty RJ 1996 PGE2 modulates the tetrodotoxin-resistant sodium current in neonatal rat DRG neurones via the cAMP-protein kinase A cascade. J Physiol 495:429–440

Fitzgerald EM, Okuse K, Wood JN, Dolphin AC, Moss SJ 1999 cAMP-dependent phosphorylation of the tetrodotoxin-resistant voltage-dependent sodium channel SNS. J Physiol 516:433–446

Fjell J, Cummins TR, Davis BM et al 1999 Sodium channel expression in NGF-overexpressing transgenic mice. J Neurosci Res 57:39–47

Gold MS, Reichling DB, Shuster MJ, Levine JD 1996 Hyperalgesic agents increase a tetrodotoxin-resistant Na current in nociceptors. Proc Nat Acad Sci USA 93:1108–1112

Gold MS 1999 Tetrodotoxin-resistant Na+ currents and inflammatory hyperalgesia. Proc Natl Acad Sci USA 96:7645–7649

Grosskreutz J, Quasthoff S, Kühn M, Grafe P 1996 Capsaicin blocks tetrodotoxin-resistant sodium potentials and calcium potentials in unmyelinated C fibres of biopsied human sural nerve *in vitro*. Neurosci Lett 208:49–52

Hartshorne RP, Messner DJ, Coppersmith JC, Catterall WA 1982 The saxitoxin receptor of the sodium channel from rat brain. Evidence for two nonidentical beta subunits. J Biol Chem 257:13888–13891

Isom LL 2000 Cellular and molecular biology of sodium channel beta-subunits: therapeutic implications for pain? Am J Physiol Gastrointest Liver Physiol 278:G349–G353

Isom 2002 β subunits: players in neuronal hyperexcitability? In: Sodium channels and neuronal hyperexcitability. Wiley, Chichester (Novartis Found Symp 241) p 124–143

Kazen-Gillespie KA, Ragsdale DS, D'Andrea MR, Mattei LN, Rogers KE, Isom LL 2000 Cloning, localization, and functional expression of sodium channel beta1A subunits. J Biol Chem 275:1079–1088

Khasar SG, Gold MS, Levine JD 1998 A tetrodotoxin-resistant sodium current mediates inflammatory pain in the rat. Neurosci Lett 256:17–20

Kral MG, Xiong Z, Study RE 1999 Alteration of Na+ currents in dorsal root ganglion neurons from rats with a painful neuropathy. Pain 81:15–24

Le Y, Sauer B 2000 Conditional gene knockout using cre recombinase. Methods Mol Biol 136:477–485

Lewin GR, Ritter AM, Mendell LM 1993 Nerve growth factor-induced hyperalgesia in the neonatal and adult rat. J Neurosci 13:2136–1248

Lyu YS, Park SK, Chung K, Chung JM 2000 Low dose of tetrodotoxin reduces neuropathic pain behaviors in an animal model. Brain Res 871:98–103

Malhotra JD, Kazen-Gillespie K, Hortsch M, Isom LL 2000 Sodium channel beta subunits mediate homophilic cell adhesion and recruit ankyrin to points of cell-cell contact. J Biol Chem 275:11383–11388

Marshall KR, Lachmann RH, Efstathiou S, Rinaldi A, Preston CM 2000 Long-term transgene expression in mice infected with a herpes simplex virus type 1 mutant severely impaired for immediate-early gene expression. J Virol 74:956–964

Matsuda Y, Yoshida S, Yonezawa T 1978 Tetrodotoxin sensitivity and Ca component of action potentials of mouse dorsal root ganglion cells cultured *in vitro*. Brain Res 154:69–82

Morgan K, Stevens EB, Shah B et al 2000 AP beta 3: an additional auxiliary subunit of the voltage-sensitive sodium channel that modulates channel gating with distinct kinetics. Proc Natl Acad Sci USA 97:2308–2313

Ogata N, Tatebayashi H 1992 Ontogenic development of the TTX-sensitive and TTX-insensitive Na channels in neurons of the rat dorsal root ganglia. Brain Res Dev Brain Res 65:93–100

Planells-Cases R, Caprini M, Zhang J et al 2000 Neuronal death and perinatal lethality in voltage-gated sodium channel alpha(II)-deficient mice. Biophys J 78:2878–2891

Ratcliffe CF, Qu Y, McCormick KA et al 2000 A sodium channel signaling complex: modulation by associated receptor protein tyrosine phosphatase beta. Nat Neurosci 3:437–444

Rinaldi A, Marshall KR, Preston CM 1999 A non-cytotoxic herpes simplex virus vector which expresses Cre recombinase directs efficient site specific recombination. Virus Res 65:11–20

Roy ML, Narahashi T 1992 Differential properties of TTX-sensitive and TTX-resistant sodium currents in rat dorsal root ganglion neurons. J Neurosci 12:2104–2111

Rush AM, Brau ME, Elliott AA, Elliott JR 1998 Electrophysiological properties of sodium current subtypes in small cells from adult rat dorsal root ganglia. J Physiol 511:771–789

Strassman AM, Raymond SA 1999 Electrophysiological evidence for tetrodotoxin-resistant sodium channels in slowly conducting dural sensory fibers. J Neurophysiol 81:413–424

Watanabe E, Fujikawa A, Matsunaga H et al 2000 Nav2/NaG channel is involved in control of salt-intake behavior in the CNS. J Neurosci 20:7743–7751

Waxman SG, Kocsis JD, Black JA 1994 Type III sodium channel mRNA is expressed in embryonic but not adult spinal sensory neurons, and is re-expressed following axotomy. J Neurophysiol 72:466–470

Waxman SG, Dib-Hajj S, Cummins TR, Black JA 1999 Sodium channels and pain. Proc Natl Acad Sci USA 96:7635–7639

Zur KB, Oh Y, Waxman SG, Black JA 1995 Differential up-regulation of sodium channel alpha- and beta 1-subunit mRNAs in cultured embryonic DRG neurons following exposure to NGF. Brain Res Mol Brain Res 30:97–105

DISCUSSION

Cummins: GDNF is certainly interesting with regard to neuropathic pain, but it is not just the type 3 channel that is regulated by GDNF. For example, we see more selective regulation of the NaN-type current with GDNF than we do with NGF. There is probably a multiplicity of factors that GDNF regulates and NGF doesn't.

Wood: I should say that we do see an up-regulation of SNS in adjacent ganglia in some models of neuropathic pain. I know other people don't see this.

Strichartz: That is important, and it is consistent with observations from the Johns Hopkins group using spinal nerve ligation in monkeys. They have shown that hyperactivity in C fibres emanating from uninjured ganglia were elevated. Isn't this consistent with Porecca's hypothesis that elevation of PN3 is important for mechanical allodynia?

Gold: I have a methodological question. My recollection is that the Waxman group has shown an NGF-induced suppression of brain 3 expression.

Waxman: Yes, we showed an up-regulation of SNS and a down-regulation of the type 3 channel in an *in vitro* model of axotomy.

Gold: In the *in vivo* experiment with an NGF cuff, did you look for brain 3 expression? I thought you got a recovery of SNS *in vivo*.

Cummins: In the NGF cuff paper (Dib-Hajj et al 1998) we didn't look at type 3 expression, we just looked at SNS.

Waxman: On the other hand, *in vitro*, we did see a modest effect of NGF in up-regulating 3.

Wood: Of course, in these *in vivo* experiments we never know how direct the effects are, so it is quite difficult to compare with the *in vitro* studies.

Bevan: With regard to the *in vitro* effects of NGF, perhaps you can clarify things. Did you say that the SNS transcript didn't change with NGF, but that it was a splice variant?

Wood: We saw a very modest up-regulation. If you remove all NGF for a long period, you can down-regulate the level of expression, but there seems to be a tonic level of SNS message.

Bevan: So you will get a down-regulation on removal of NGF.

Wood: There is some debate about this. There certainly is a small increase in immunoreactive protein in response to NGF.

Bevan: In terms of the functional currents, it is very clear that if you remove the NGF you lose the TTX-R SNS-type currents.

Wood: In inflammatory pain states, I am not sure whether the alterations in the level of NGF are adequate to increase functional levels of TTX-R currents.

Strichartz: Are these TrkA (high affinity) receptor-activating levels of NGF, or might they also get p75 (low affinity)?

Wood: In these *in vivo* experiments very large amounts of material are used because of the problem of circulation and degradation, so p75 could be activated.

Strichartz: What about *in vitro*?

Wood: We tend to use nanogram concentrations, which are on the borderline of acting on p75.

Gold: My interpretation of the data that you presented would be that you have come down on the side of brain 3 underlying the neuropathic pain, as opposed to SNS. Could you lay out your arguments?

Wood: There is a very good literature for ectopic action potential propagation that is TTX-S, which you would not expect if SNS underlay that phenomenon.

Gold: The problem with the ectopic activity experiment is that it doesn't look like SNS in the axon really can produce conduction. You sort of said that in the beginning, but for instance in the experiments by Brock et al (1998) they can record TTX-R spike generation in the cornea, but they are still blocking conduction with TTX in those axons. Everyone else who has looked in axons can block conduction in axons. Even though Grafe's group has demonstrated that there are functional TTX-R currents in axons, I don't believe that anyone has failed to block conduction.

Wood: I think this may depend on the nerve. If you look at rat dural membrane, you certainly can block a large percentage of C fibres in the presence of TTX.

Gold: My recollection is that TTX was applied to the dura. If so, then the question is one of the ability of TTX to penetrate the dura.

Wood: But the A fibre component was completely ablated.

Gold: Then it is an issue of the sensitivity of spike initiation and propogation *and* the ability of TTX to penetrate the dura.

Wood: The argument is that SNS is actually present at high levels at the site of nerve injury, and yet we find the mRNA goes away. These experiments depend on the use of antisera which are not necessarily well characterized. Obviously, when you transect a nerve, everything piles up because of axonal transport. I don't think this necessarily has any functional significance. It is hard to believe that you are getting increased protein expression when the message has gone away.

Gold: My problem is in interpreting the ectopic activity as an underlying mechanism, given that there are now conflicting results in the literature from people who have been unable to eliminate the pain behaviour by eliminating the ectopic activity, or at least the source of ectopic activity. This brings into question what the underlying mechanism is. I am not saying that there can't be an increase in excitability that requires an up-regulation of brain 3, but I still question the function of the ectopic activity, and whether this alone can be used to dismiss the function of SNS.

Wood: It is completely possible that the GDNF therapeutic effects are not mediated in any way by the effects on Na^+ channels. This remains a formal possibility. Until we can actually do the ablation experiment with the type 3 channel, we can't answer that question.

Waxman: At least this helps us begin to dissect out the roles of all of the various channels.

Cummins: One of the hypotheses with the type 3 channel is that it may be faster at repriming than the normal currents in these neurons. Certainly, we see a shift in repriming with injury in some neurons, and a dramatic shift in the small neurons. Recently we have been looking at the type 3 channels expressed in HEK cells. It does reprime faster than the PN1 channel, for example. This faster repriming would be expected to enhance the ability of these neurons to fire repetitively. This is how type 3 could be involved.

Segal: There are a variety of clinical syndromes with congenital insensitivity to pain. Have people looked at these for channel defects?

Wood: The only well characterized ones that I am aware of involve TrkA. To my knowledge none of the channels we have been talking about today have been implicated in insensitivity to pain syndromes.

Bean: What is the state of knowledge about the SNSA transcript? Do you think it makes channels?

Wood: We got very excited, so we reinjected it into the knockout mice, which we thought would be a perfect expression system, but it didn't work. But there were also problems expressing the NaN channel. We still have some mysteries to resolve.

Bevan: If you look at the inflammatory hyperalgesia of the SNS knockouts, how does this fit in with the antisense experiments?

Wood: We did one experiment on the effects of prostaglandin E2 on the knockout. To our disappointment, there was no obvious deficit.

Bevan: In terms of the hyperalgesia, I thought you delayed the onset of this.

Wood: Only in the presence of lidocaine did we see a dramatic inhibition of the development of inflammatory pain, which we ascribe to the compensatory expression of PN1.

Bevan: The antisense experiments looked very convincing.

Gold: I have always argued that differences between antisense and knockout data are due to differences in compensation. If you have a mouse that has grown to adulthood in the absence of a specific protein, then I would suggest that all bets are off. I have looked at neurons *in vitro*, taken from animals treated with antisense oligonucleotides *in vivo* and I was unable to detect a change in TTX-S Na^+ current. Western blot analysis would suggest that there is no change in PN1 either. While not conclusive, these data argue against compensation in antisense studies. In contrast, there is evidence for compensatory changes in TTX-S current in knockout mice.

Wood: I thought it would be important to use the SNS antisense nucleotide in the SNS knockout.

Strichartz: I was thinking of Wayne Crill's paper (Crill et al 2002, this volume) in light of our propensity to think about Na^+ channels and spike activity, and looking for increasing spike activity as a basis for pathologies associated with Na^+ channels. Perhaps some of these aberrant Na^+ channels, whichever ones they happen to be, are inserting in membranes at points where it is not spike activity per se but coupling to generator potentials or facilitation of release presynaptically. We haven't really looked at the terminals (in fact it's almost impossible to do this) but there may be some interesting relationships there, when we can look at the pathophysiology of the endings of these very small nerves.

Waxman: We know from immunocytochemistry that type 3 channels are shipped to the injured axonal tips (Black et al 1999). Whether they are inserted into the membrane and can function there is not clear. But you are right: in addition to thinking about the rising phase of the spike, we have to think about how the cell gets there and the entire transductive machinery. This is especially the case in cells that are endowed with channels whose activation curves are shifted so far to the left, one wonders about subthreshold roles of at least some of these channels.

Bevan: In the context of terminals, obviously one of the things they will do in the sensory neurons is release transmitters in the periphery. I don't know what role TTX-R currents might have in that sort of antidromic effect. Do you see any change in neurogenic inflammation in the SNS knockout mice?

Wood: We haven't done that yet.

Waxman: There is a nagging issue of the apparently different results from different laboratories. One does an experiment as simple as cutting the sciatic nerve and looking for neuropathic hyperexcitability, but the results are very different from different laboratories. It is often because of different strains of rats or mice and different operators: a chronic constriction injury in one lab may involve a tighter suture than in another laboratory, or there may be polymorphisms that hold some clues for us. It is important for us as investigators to look at the disparities and see whether we can even use them as tools.

Strichartz: There is something even more troubling that that. Zvi Seltzer has reported that his partial sectioning autotomy phenotype disappeared 'spontaneously' at some point (Shir et al 1997). They traced this variation back to the feed! The feed suppliers had been changed. There are all these uncontrolled exogenous factors which can completely change the experiment, and often we have no knowledge of them.

Horn: What is the role of antidromic activity in these pain fibres?

Wood: It is very important. It does things such as regulate bloodflow.

Horn: Is there a lot of antidromic activity in pain neurons?

Gold: The Galveston group would argue that there is. They have put together a nice story following spikes that invade the dorsal horn (Sluka et al 1995). The spikes evoke a GABA depolarization of presynaptic terminals resulting in the initiation of antidromic activity. They also looked in the knee joint in the presence of inflammation. One of the nerves innervating the knee joint is primarily afferent, yet if they cut the nerve distal to the site of recording, they were able to detect a good deal of activity. They have gone on to show, through microdialysis of inflamed tissue, the importance of the peripheral release of transmitters from the primary afferent neuron to the development of the inflammatory response to injury.

References

Black JA, Cummins TR, Plumpton C et al 1999 Upregulation of a silent sodium channel after peripheral, but not central, nerve injury in DRG neurons. J Neurophysiol 82:2776–2785

Brock JA, McLachlan EM, Belmonte C 1998 Tetrodotoxin-resistant impulses in single nociceptor nerve terminals in guinea-pig cornea. J Physiol 512:211–217

Crill WE, Schwindt PC, Oakley JC 2002 Enhanced transmission of glutamate current flowing from the dendrite to the soma in rat neocortical layer 5 neurons. In: Sodium channels and neuronal hyperexcitability. Wiley, Chichester (Novartis Found Symp 241) p 61–71

Dib-Hajj SD, Black JA, Cummins TR, Kenney AM, Kocsis JD, Waxman SD 1998 Rescue of α-SNS sodium channel expression in small dorsal root ganglion neurons after axotomy by nerve growth factor *in vitro*. J Neurophysiol 79:2668–2676

Shir Y, Ratner A, Seltzer Z 1997 Diet can modify autotomy behavior in rats following peripheral neurectomy. Neurosci Lett 236:71–74

Sluka KA, Willis WD, Westlund KN 1995 The role of dorsal root reflexes in neurogenic inflammation. Pain Forum 4:141–149

Sodium channels and epilepsy electrophysiology

Michael M. Segal

Harvard Medical School and Brigham & Women's Hospital, Longwood Medical Research Center, 221 Longwood Ave, Boston MA 02115, USA

Abstract. We examined the electrophysiology of epilepsy in the simplest system that exhibits epileptiform activity: microisland cultures that contain only one neuron. Some of these solitary excitatory hippocampal neurons generate the 'ictal' epileptiform activity characteristic of seizures. These neurons have endogenous (non-transmitter-mediated) bursts of activity that last for many seconds and appear to be driven by a persistent Na^+ current. We examined this persistent Na^+ current at the single channel level by recording the late openings of Na^+ channels using outside-out patch recordings. Phenytoin reduced the probability of these late channel openings, but had less effect on the early channel openings that make up the peak Na^+ current. The reduction of late channel openings was larger with pulses to more depolarized voltages. In contrast, the effect on early channel openings was similar at all voltages. There was little effect of phenytoin on the duration of channel openings and no effect on open channel current. This suggests that the persistent Na^+ current is crucial in generating seizures. A good strategy for selecting anticonvulsants may be to search for drugs that more selectively block the persistent Na^+ current at depolarized voltages. Such drugs could combine effectiveness and reduced side effects.

2002 Sodium channels and neuronal hyperexcitability. Wiley, Chichester (Novartis Foundation Symposium 241) p 173–188

Epilepsy is large group of different diseases characterized by one common property: chronic seizures. Hundreds of different types of epilepsy are already known. Each seizure is in itself complicated, with many currents activated simultaneously in many neurons, often affecting nearly all the neurons in the brain. This complexity of epilepsy presents great challenges for reductionistic studies. Nevertheless, progress is being made in a variety of ways. Genetic approaches have begun to whittle away at the estimated 1000 forms of genetic epilepsy, as described elsewhere in this volume. We have used another type of reductionism, developing simple neuronal circuits that display the characteristic electrophysiology of epilepsy. This work suggests that persistent Na^+ currents have a central role in the abnormal activity that produces epilepsy. The work also gives us new approaches for working with the molecule that is already the most productive target for epilepsy drugs: the neuronal Na^+ channel.

Epilepsy is typically studied clinically using the electroencephalogram (EEG), a recording of electrical field changes produced by neuronal activity. The foundation for simplifying such activity down to simpler circuits was laid by identification of the intracellular electrophysiology that underlies the EEG (Kandel & Spencer 1961, Matsumoto & Ajmone Marsan 1964). The counterpart to an EEG 'spike' was found to be an event termed a 'paroxysmal depolarizing shift' (PDS) that lasts 50–100 ms and occurs in a synchronized way in many neurons in the brain. During a seizure, there is a train of EEG 'spikes' lasting seconds to minutes, corresponding inside neurons to a train of PDSs with a sustained depolarization of the neuronal voltage. These runs of PDSs/EEG spikes are termed 'ictal' activity, while isolated PDSs/EEG spikes are termed 'interictal' activity.

The ability to recognize ictal and interictal activity in a single neuron made it possible to look for such epileptiform activity in neurons in very small circuits. Unfortunately, hopes of studying such tiny circuits were dashed by the finding that the interictal activity disappeared when a hippocampal slice was cut down to ~1000 neurons (Miles et al 1984) and ictal activity was seen only in slices that were much larger. However, both ictal and interictal epileptiform activity could be produced in cultures of ~1000 neurons (Furshpan & Potter 1989), suggesting the possibility of studying epileptiform activity in smaller circuits in culture. The problem was producing such small circuits. A strategy of using low density cultures that works with invertebrate neurons (Schacher et al 1985) was difficult to use with vertebrate neurons because such neurons have poor survival in low density cultures. Using small microisland cultures, as was done for studies of vertebrate autonomic neurons (Furshpan et al 1976, 1986), was scuttled by the tendency of central neurons to be accompanied by glial cells, which stuck to the substrate and produced a confluent mass culture instead of many microisland cultures. Success was finally achieved using a thin film of agarose as a material that is non-adhesive to glial cells. The agarose provides a 'sea' separating microislands composed of substrate supporting small circuits of central neurons (Segal & Furshpan 1990, *http://segal.org/microislands*).

Small circuits of hippocampal neurons grown in microisland cultures displayed both ictal and interictal epileptiform activity. Not only was both ictal and interictal activity seen in circuits with two excitatory neurons, but both types of activity were seen also in two-neuron circuits containing one excitatory neuron and one inhibitory neuron. The activity in these 'EI' microislands persisted even when inhibitory transmission was blocked pharmacologically, suggesting that epileptiform activity could be generated under these conditions by the single excitatory neuron. Studies of such solitary neurons produced the simplest neuronal system displaying epileptiform activity: a solitary excitatory neuron making synaptic contact on itself in a microisland culture (Segal 1991).

Clearly the solitary neuron model is just an abstraction of much larger circuits in the brain. But such reductionism affords not only a conceptual model for approaching epilepsy but also an experimental model for testing certain hypotheses. The one neuron model appears to be the minimal circuit for cortical forms of epilepsy, but for modelling subcortical 'absence' epilepsy it appears that a two neuron model is the minimal circuit: one excitatory neuron with prominent Ca^{2+} currents connected to one inhibitory neuron with prominent Ca^{2+} currents (von Krosigk et al 1993).

We have analysed the epileptiform activity in the solitary neuron model of cortical epilepsy in order to gain insights into the nature of ictal and interictal activity. Every solitary excitatory neuron that we tested displayed some epileptiform activity. Some neurons displayed only isolated interictal PDSs, while others also displayed runs of PDSs with sustained depolarizations characteristic of ictal activity (Segal 1991). The reason for the difference between these two types of neurons was an intriguing question. Finding the distinctive currents of the neurons with interictal activity could offer insights into the currents responsible for interictal events progressing into ictal events. The interictal events have been well characterized (Johnston & Brown 1981), but clinicians know that interictal activity is not disruptive to consciousness, and anticonvulsants leave interictal activity undisturbed (Gotman & Marciani 1985). The goal in epilepsy therapy is to prevent or shorten the runs of events that constitute ictal activity. However, ictal activity has been difficult to study previously due to the complexity of circuits displaying epileptiform activity.

Comparison of solitary neurons with ictal versus interictal activity revealed that the neurons with ictal activity retained their distinctiveness even after all synaptic transmission was blocked using a 'synaptic blocking solution' in which Ca^{2+} was replaced by Mg^{2+} and the glutamate antagonists APV and CNQX were added (Segal 1994). The neurons with ictal activity in regular solutions displayed endogenous bursts of electrical activity that lasted for many seconds in the synaptic blocking solution. These bursts had long plateau depolarizations to ~ -20 mV with many action potentials generated during the plateau depolarization. In contrast, neurons with only interictal activity in regular solutions displayed one or several action potentials in the synaptic blocking solution. This suggested that the vigorous endogenous activity with plateau depolarizations was responsible for transforming isolated PDSs recorded in regular solutions into longer-lasting events with runs of PDSs with sustained depolarizations of the neuronal voltage. It was also clear that there was a persistent Na^+ current in these neurons because introducing tetrodotoxin in the middle of a pacemaker potential leading to a burst aborted the depolarization as well as preventing the burst (Segal 1994).

Since persistent Na^+ currents are small compared to peak Na^+ currents, we have studied the Na^+ currents at the single-channel level, allowing for unambiguous, though laborious determinations of Na^+ currents without concerns about artefacts due to small whole-cell currents or poor voltage clamp control. We found late openings of Na^+ channels using both cell-attached patch recording and outside–out patch recording (Segal & Douglas 1997) similar to those recorded from layer 5 neocortical neurons in slices (Alzheimer et al 1993). Although the cell-attached patch recordings of Na^+ channels with intact intracellular environment are more physiological (Aldrich & Stevens 1987), this configuration was not analysed in detail due to the difficulty of eliminating K^+ currents. The Na^+ channel openings in outside–out patches were analysed in detail, although their open times were more prolonged than those seen in the more physiological cell-attached patches. Using both cell-attached patches and outside–out patches we found that all inward currents were blocked by tetrodotoxin. Using outside–out patches we found that all inward channel currents were of uniform amplitude with 14.3 pS chord conductance and an extrapolated reversal potential far above 0 mV; no evidence of cation channels was seen. All inward currents were blocked by tetrodotoxin (1 μM) and all outward currents were blocked by the caesium in the recording electrode, suggesting that the Na^+ current is the major endogenous inward current in these neurons.

All patches with Na^+ channels had both early and late channel openings, corresponding to peak and persistent Na^+ currents. The persistent Na^+ current was typically 0.5% of the peak current. The Na^+ channel openings included both the 'brief late openings' and the burst openings described in other neuronal (Alzheimer et al 1993) and muscle cells (Patlak & Ortiz 1986), openings that are believed to represent different kinetic states of the same channel molecule (Moorman et al 1990). The burst openings were seen in far less than 1% of pulses, and in some patches were not seen at all.

From these experiments it appears that the inward current that drives the neuronal bursts that underlie the ictal activity is a persistent Na^+ current. We were not able to get large numbers of experiments in which whole cell activity could be correlated with single channel events from the same neuron. Therefore it is not clear whether the endogenous bursting in the neurons with ictal activity is due to an increase in overall persistent Na^+ current, a specific increase in burst openings, or a decrease in countervailing K^+ currents.

It should not have come as a surprise to find that Na^+ currents play a key role in generating seizures since many of the best anticonvulsants act by blocking Na^+ channels (Rogawski & Porter 1990). This anticonvulsant mechanism of Na^+ channel antagonists had been posited previously as being due to the ability of the antagonists to block 'sustained repetitive firing' of neurons—i.e. a drug effect on peak Na^+ currents that blocked the spread of seizure activity by action potentials travelling along axons (McLean & Macdonald 1986, Macdonald 1988). To test this model, we examined the effects of phenytoin on ictal activity in solitary excitatory neurons in microisland cultures. In this system, transmission of epileptiform activity to other neurons is irrelevant since there are no other neurons in the circuit.

We found an anticonvulsant effect of phenytoin in microisland cultures with a solitary excitatory neuron (Segal 1994). Plateau depolarization events in synaptic blocking solution were abolished by therapeutic levels of phenytoin (8 μM), with the plateau depolarization being replaced by a burst of action potentials. These bursts of action potentials were not affected by therapeutic levels of phenytoin. In contrast to this large 'safety factor' for action potential generation, there was no safety factor for the plateau depolarizations underlying ictal activity, and these ictal events were highly sensitive to phenytoin, mirroring the anti-ictal efficacy of anticonvulsants seen clinically and the lack of such an interictal effect (Gotman & Koffler 1989). This work suggests that phenytoin acts through its effect on the persistent Na^+ current that drives the plateau depolarizations underlying epileptic bursts. The 'sustained repetitive firing' used in previous models (McLean & Macdonald 1986, Macdonald 1988) was produced by depolarizing the neuron with injected current, so it is likely that the effect observed in that model was due to simulating the plateau depolarization rather than an effect on sustained repetitive firing in itself. These sustained firing and Na^+ current mechanisms can be separated well using the microisland model.

Previous studies of phenytoin action have examined the effects of the drug on currents in whole neurons (Lang et al 1993, Kuo & Bean 1994) or on currents through proteolytically modified channels (Quandt 1988). We explored the effects of phenytoin directly on native channels generating the persistent Na^+ current that appears to underlie the ictal activity. Phenytoin has much more effect on the late openings of Na^+ channels underlying the persistent Na^+ current than it does on the early channel openings underlying the peak Na^+ current (Segal & Douglas 1997). This selective effect of phenytoin for persistent Na^+ current was highly dependent on the voltage of the test pulse. With test pulses to -50 mV, phenytoin (60 μM) reduced late and early currents equally (to ~40% of control) but with test pulses to 0 mV the drug reduced late currents to ~13%, while still reducing early currents to ~40% of control. The effect of phenytoin was almost exclusively via reduction in the number of channel openings; there was little effect of phenytoin on the duration of channel openings and no effect on open channel current. This is consistent with models in which phenytoin binds to an inactivated state of the Na^+ channel (Kuo & Bean 1994). Thus, phenytoin is quite selective for antagonizing persistent Na^+ currents at depolarized voltages, properties that may be ideal for anticonvulsant action since ictal activity appears to depend on a persistent Na^+ current that holds a neuron at a depolarized potential.

This selectivity of phenytoin for persistent Na^+ currents at depolarized potentials raises the hope of finding improved anticonvulsant drugs. Drugs such as phenytoin produce mild sedation, a disabling side effect (Rogawski & Porter 1990). If the sedation is produced by the effect of phenytoin on the peak Na^+ current, while the anticonvulsant effect is produced by the effects on persistent Na^+ current, it may be possible to separate sedation from therapeutic drug effect by finding a drug that is more selective for the persistent Na^+ current than is phenytoin. Alternatively, it is possible that sedation and the therapeutic action are inextricably linked since sedation could result from a reduction of the duration or frequency of normal neuronal bursting. Since neuronal bursting is important in development of the nervous system it is also possible that teratogenic effects of phenytoin could be due to reducing neuronal bursting. However in the only well-studied example of such neuronal bursting in development, the retinal bursting needed to form proper retino-tectal connections, the bursting is adenosine-mediated and insensitive to tetrodotoxin (Stellwagen et al 1999).

Using the single-channel studies we examined a drug that was a candidate to be such an improved Na^+-channel anticonvulsant, memantine (Meldrum et al 1986). Memantine has been used in humans for many years and has negligible side effects (Chen et al 1998). Using the 'sustained repetitive firing' model it was concluded that the anticonvulsant action of memantine was due to an effect on voltage-sensitive Na^+ channels (McLean 1987, Netzer & Bigalke 1990). However, this conclusion was difficult to rely on since sustained repetitive firing could also be produced by effects on transmitter channels. Indeed, Chen & Lipton (1997) found that memantine is an uncompetitive NMDA antagonist. To test the mechanism of memantine action and the predictive powers of the sustained repetitive firing model we examined the effect of memantine directly by using the single channel model. We did not find appreciable blockade of Na^+ currents (Lundquist 1999). Therefore, it is likely that the anticonvulsant action of memantine is due to NMDA antagonism and not to a Na^+ channel effect. Furthermore inhibition of sustained repetitive firing model can no longer be considered a reliable indicator of action at the Na^+ channel. The direct Na^+ channel techniques give more reliable and detailed information about the effect of a drug on peak and sustained Na^+ currents.

In summary, it appears that the persistent Na^+ current has a major role in driving the ictal activity that underlies seizures. Although mutations of Na^+ channels are likely to underlie only a small fraction of cases of epilepsy, it is likely that the persistent Na^+ current represents a final common pathway in many cases of epilepsy. A good strategy for developing new anticonvulsants with fewer side effects may be to search for drugs that more selectively block the persistent Na^+ current at depolarized voltages.

Acknowledgements

This work was supported by the Klingenstein Fund, the Dana Foundation and by NIH (NINDS).

References

Aldrich RW, Stevens CF 1987 Voltage-dependent gating of single sodium channels from mammalian neuroblastoma cells. J Neurosci 7:418–431

Alzheimer C, Schwindt PC, Crill WE 1993 Modal gating of Na^+ channels as a mechanism of persistent Na^+ current in pyramidal neurons from rat and cat sensorimotor cortex. J Neurosci 13:660–673

Chen HS, Lipton SA 1997 Mechanism of memantine block of NMDA-activated channels in rat retinal ganglion cells: uncompetitive antagonism. J Physiol 499:27–46

Chen HSV, Wang YF, Rayudu PV et al 1998 Neuroprotective concentrations of the N-methyl-D-aspartate open-channel blocker memantine are effective without cytoplasmic vacuolation following post-ischemic administration and do not block maze learning or long-term potentiation. Neuroscience 86:1121–1132

Furshpan EJ, Potter DD 1989 Seizure-like activity and cellular damage in rat hippocampal neurons in cell culture. Neuron 3:199–207

Furshpan EJ, MacLeish PR, O'Lague PH, Potter DD 1976 Chemical transmission between rat sympathetic neurons and cardiac myocytes developing in microcultures: evidence for cholinergic, adrenergic, and dual function neurons. Proc Natl Acad Sci USA 73:4225–4229

Furshpan EJ, Landis SC, Matsumoto SG, Potter DD 1986 Synaptic functions in rat sympathetic neurons in microcultures. I. Secretion of norepinephrine and acetylcholine. J Neurosci 6:1061–1079

Gotman J, Marciani MG 1985 Electroencephalographic spiking activity, drug levels, and seizure occurrence in epileptic patients. Ann Neurol 17:597–603

Johnston D, Brown TH 1981 Giant synaptic potential hypothesis for epileptiform activity. Science 211:294–297

Kandel ER, Spencer WA 1961 Excitation and inhibition of single pyramidal cells during hippocampal seizure. Exp Neurol 4:162–179

Kuo CC, Bean BP 1994 Slow binding of phenytoin to inactivated sodium channels in rat hippocampal neurons. Mol Pharmacol 46:716–725

Lang DG, Wang CM, Cooper BR 1993 Lamotrigine, phenytoin and carbamazepine interactions on the sodium current present in N4TG1 mouse neuroblastoma cells. J Pharmacol Exp Therap 266:829–835

Lundquist ER 1999 Memantine as a possible anti-epileptic drug: examining its mechanism of action using models that test sodium current. Undergraduate thesis, Harvard University, Cambridge, MA, USA

Macdonald RL 1988 Anticonvulsant drug actions on neurons in cell culture J Neural Transm 72:173–183

Matsumoto H, Ajmone Marsan C 1964 Cortical cellular phenomena in experimental epilepsy: Ictal manifestations. Exp Neurol 9:305–326

McLean MJ 1987 *In vitro* electrophysiological evidence predicting anticonvulsant efficacy of memantine and flunarizine. Pol J Pharmacol Pharm 39:513–525

McLean MJ, Macdonald RL 1986 Carbamazepine and 10,11-Epoxycarbamazepine produce use- and voltage-dependent limitation of rapidly firing action potentials of mouse central neurons in cell culture. J Pharm Exp Therap 238:727–738

Meldrum BS, Turski L, Schwarz M, Czuczwar SJ, Sontag KH 1986 Anticonvulsant action of 1,3-dimethyl-5-aminoadamantane. Pharmacological studies in rodents and baboon, Papio papio. Naunyn Schmiedebergs Arch Pharmacol 332:93–97
Miles R, Wong RKS, Traub RD 1984 Synchronized afterdischarges in the hippocampus: Contribution of local synaptic interactions. Neuroscience 12:1179–1189
Moorman JR, Kirsch GE, VanDongen AMJ, Joho RH, Brown AM 1990 Fast and slow gating of sodium channels encoded by a single mRNA. Neuron 4:243–252
Netzer R, Bigalke H 1990 Memantine reduces repetitive action potential firing in spinal cord nerve cell cultures. Eur J Pharmacol 186:149–155
Patlak JB, Ortiz M 1986 Two modes of gating during late Na^+ channel currents in frog sartorius muscle. J Gen Physiol 87:305–326
Quandt FN 1988 Modification of slow inactivation of single sodium channels by phenytoin in neuroblastoma cells. Mol Pharmacol 34:557–565
Rogawski MA, Porter RJ 1990 Antiepileptic drugs: pharmacological mechanisms and clinical efficacy with consideration of promising developmental stage compounds. Pharmacol Rev 42:223–286
Schacher S, Rayport SG, Ambron RT 1985 Giant Aplysia neuron R2 reliably forms strong chemical connections *in vitro*. J Neurosci 5:2851–2856
Segal MM 1991 Epileptiform activity in microcultures containing one excitatory hippocampal neuron. J Neurophysiol 65:761–770
Segal MM 1994 Endogenous bursts underlie seizure-like activity in solitary excitatory hippocampal neurons in microcultures. J Neurophysiol 72:1874–1884
Segal MM, Douglas AF 1997 Late sodium channel openings underlying epileptiform activity are preferentially diminished by the anticonvulsant phenytoin. J Neurophysiol 77:3021–3034
Segal MM, Furshpan EJ 1990 Epileptiform activity in microcultures containing small numbers of hippocampal neurons. J Neurophysiol 64:1390–1399
Stellwagen D, Shatz CJ, Feller MB 1999 Dynamics of retinal waves are controlled by cyclic AMP. Neuron 24:673–685
von Krosigk M, Bal T, McCormick DA 1993 Cellular mechanisms of a synchronized oscillation in the thalamus. Science 261:361–364

DISCUSSION

Horn: When you pull your patches you have a heterogeneous population of channels. Presumably you don't think you are looking at one type of Na^+ channel.

Segal: All our recordings have between 10 and 50 openings in the peak current, so it is difficult to answer your question.

Horn: But are they different isoforms? Can you see different single channel amplitudes?

Segal: All the amplitudes are the same. We didn't find anything that demonstrated the existence of different types of Na^+ channel in the neurons we studied. The best evidence about the number of channel types required to produce the different types of channel openings is the finding that expressing a single mRNA species produces several types of channel openings (Moorman et al 1990). This suggests that all the events we recorded could arise from the same type of channel in different kinetic states or with different covalent modulation.

Horn: You have these very highly active long-duration bursts. You exclude them from your analysis, but it seems that if you just integrate the total amount of time spent open, there is much more in those than in the absence of the bursts.

Segal: We analysed it both ways and the results were similar. The only reason we tried analysing without the long bursts was to make sure that our results were not dependent on chance occurrences of clusters of rare, long bursts, which could occur by chance.

Raman: I have a more general question that comes from your use of the term 'persistent Na^+ current'. We all use this term, and I am interested in some formal definition of what this really refers to. Sometimes we define a persistent Na^+ current as the current that is active a few milliseconds after a step depolarization, and sometimes as the current active a few *seconds* after depolarization. Is the relevant feature here the open probability of one channel near −60 mV before an action potential occurs, but then that so-called persistent current may inactivate during the spike? Or is it that one channel occasionally goes into a non-inactivating mode? Does anyone still believe in a truly non-inactivating, TTX-sensitive Na^+ channel that is physically separate from the other Na^+ channels?

Segal: There are TTX-resistant Na^+ channels that have persistent currents, but the Na^+ currents in our system are completely blocked by TTX. The only evidence that seems clear on the question of multiple channel types is that a single mRNA can produce Na^+ channels displaying the different types of channel openings (Moorman et al 1990).

Raman: What do you have in mind when you say that it will be interesting to find drugs that will target specifically persistent Na^+ current? Is it simply going to be something that stabilizes inactivated states?

Segal: That would do it. However, if there is some type of modification of a channel, such as phosphorylation, that changes the late channel openings, that would be another target for pharmacological intervention.

Horn: Under the condition of your experiments there is no phosphorylation/dephosphorylation going on.

Segal: Presumably not, although I am continually surprised by reports of how much is left attached to the membrane and how much can go on under these conditions.

Goldin: Are there any differences between the delayed openings and the early openings in terms of open time or amplitude?

Segal: There are no differences in amplitude. Typically, the individual early channel openings making up the peak currents can be discerned, and these channel openings have the same amplitude as the delayed openings. It is difficult to comment on the open times of the early openings since a large number of channels open at once. All the analysis of what I refer to as 'late current' was from the period of 100–750 ms into the voltage pulse.

Goldin: That seems like an impressively long open time. Is that true?

Segal: Some channels do stay open in a bursting mode for that entire time. It is very impressive and may be a major source of current in pathological states.

Baker: We have seen something similar in large diameter dorsal toot ganglion (DRG) neurons (Baker & Bostock 1998). The late currents are very complicated, the channels exhibiting ultra brief openings, openings about 10 ms in duration, and very long bursts (some at least 400 ms long). None of the openings that we saw that could be contributing to the late currents were appropriate for generating the transient current in the patch. The jury is out on whether or not you have separate molecular species of channels generating transient and late currents. Laying that aside, we concluded that the channels that were giving rise to the late currents are not the same as the channels giving rise to the transient currents.

Goldin: Are they flickering when they are open later?

Baker: Yes, the channels flicker, but the flickering becomes less obvious as they become more and more depolarized. With sufficient depolarization, one can record very long openings several hundred milliseconds in length, which are like boxes. I wasn't sure whether this was due to some voltage-dependent block which became so rapid it was impossible for me to resolve it at the depolarized potential.

Cummins: One explanation for the higher affinity of the persistent current is that it is stabilizing the inactivated state and blocking the channel. An alternate explanation might not involve phenytoin actually getting into the local anaesthetic binding site that Bill Catterall's lab has identified. Instead, perhaps low levels of phenytoin are actually facilitating mode shifting and biasing the channels towards the fast mode. Maybe it is just acting through membrane fluidity or some other effect that is not the same as the mechanism for blocking the channel. Perhaps we should be looking for mode shifters and not channel blockers or inactivation state stabilizers.

Baker: In the DRG, one mode shifter appeared to be change in pH, in as much as late current is very sensitive to acidification (Baker & Bostock 1999). If you acidify, the late currents disappear. In part, this seems to be due a change in gating rather than just simply proton block. I would suggest that pH is one way in which gating mode can be modulated.

Strichartz: One technical suggestion, and one thought. If you think that you have multiple channels in a patch, it is possible to pipe UV light down to that patch with a fibre optic rod, and with single photon hits to knock out one or several channels. According to earlier work that Wolf Almers did with micropatches, in practice this should be possible (Weiss et al 1986). This is one way of trying to discriminate contributions from different proteins. With respect to modulation of gating versus pore blockade, the evidence from gating currents for local anaesthetic-like compounds is that it is not pore blockade per se, but incomplete activation of channels which accounts for most of the disappearance of currents. This shows up as a null opening for a series of sweeps looking at single channels. This is consistent with the general idea that Ted Cummins suggested that these might be modulations rather than occlusions. In fact, one sees blockade for very brief depolarizations which really only permit open channels but not inactivated ones to be around for any length of time. The dissociation and subsequent availability data from prepulses indicate that the channels are now behaving as if they are in the inactivated state. You don't need an inactivated channel to get an 'inactivated'-type stabilization — you can get that through an open channel blockade. It seems that they all degenerate down to that common state.

Catterall: One system that we have studied, the type 2 channel, which is a fast-gating, fast-inactivated brain channel, has a persistent current of about 2% or so when it is transfected into mammalian cells. This can be dramatically increased by co-transfecting G protein $\beta\psi$ subunits, creating a situation with 20% persistent current on average. So with one (admittedly unphysiological) manipulation, you can generate huge persistent currents from a Na^+ channel that we know is present in all of the cells that generate them. We haven't been able to show that activation of G protein-coupled receptors in a neuron, or even in a transfected cell, can reproduce this effect. So far this is an effect of overexpression of $G\beta\psi$. Nevertheless, I think that this is a likely mechanism in neurons. We just don't know how to stimulate the neurons to get this kind of effect. Certainly, in this case you can modulate the level of persistent current in transfected cells with an intracellular transduction protein.

Raman: When you are calling it persistent current, do you mean that there is some late steady state current with some step depolarization to some potential?

Catterall: The current is sustained as long as you depolarize.

Raman: And it will not inactivate if you depolarize to a more positive value?

Catterall: You can inactivate all of the channels if you predepolarize for a very long time. In that case you have no transient current left, and hence no persistent current.

Raman: So you would define 'persistent current' as a component of the current that passes through a channel that is capable of inactivation at a more positive potential.

Catterall: Although we haven't done much single channel recording, we think it's a gating mode change of the kind that Wayne Crill described in cortical neurons and other people have described elsewhere. The only difference in these experiments is that we have been able to make a manipulation that greatly changes the amount of persistent current. We think somehow that the G protein puts the channel in a state in which it doesn't inactivate, and we suspect this is due to direct binding of the G protein $\beta\gamma$ subunits to the channel.

From the mutagenesis experiments I will discuss in my paper, the binding of the drugs that we think are anticonvulsants seems to be affected by the same mutations in the same way as the binding of local anaesthetics. We think the same receptor site is involved in binding the anticonvulsant and local anaesthetic drugs and although binding there may also inhibit gating currents, as Gary Strichartz mentioned earlier, this drug binding probably also blocks the current.

Cummins: This doesn't rule out the possibility that phenytoin might also have an effect somewhere else as a mode shifter.

Catterall: It doesn't rule that out, but it does say that at the concentrations at which it is anticonvulsant it is binding to the site that blocks Na^+ channels.

Crill: With the great diversity if channels, one can do paper physiology and get bursts many different ways. It is quite possible that the anticonvulsants might affect the Na^+ channel currents and have an effect by decreasing excitability, but that doesn't say much about what is causing the epilepsy.

Segal: Although there is a type of epilepsy in humans caused by a Na^+ channel mutation (Wallace et al 1998) it is clear that Na^+ channel mutations will account for only a small fraction of epilepsy. Of the causes of epilepsy that are understood at the genetic level, most are due to cell migrational problems, defects in energy metabolism or accumulation of material that disrupts neuronal function. However, the situation is reversed when we look at epilepsy treatment instead of epilepsy causes. Most of the antiepileptic drugs have their major action at Na^+ channels (Rogawski & Porter 1990). Our trial and error discovery of anticonvulsant drugs has arrived at the Na^+ channel again and again since the Na^+ current is one of the final common pathways in expressing the seizure phenotype. As we learn more about the final common pathways such as the Na^+ channel it will be easier to work back and look at preceding steps involving protein kinases and other intracellular modulators.

Ptacek: I have always been struck by the fact that most anticonvulsants (perhaps all) are dirty drugs. The way I have been thinking about the tight regulation of neuronal excitability is that even though there are variations, if you have enough variables, the fit of net excitability around the mean will be tight. Perhaps the same is true when we are talking about anticonvulsants. It is aesthetically pleasing to think we could have a very selective drug, but perhaps this would have too strong an effect and toxic consequences.

Segal: One of the fascinating things about phenytoin is that it is not just a weaker TTX: it acts more on the persistent current and acts more with bigger depolarizations. One of the NMDA antagonists, memantine, also has similar properties: it hardly blocks at the beginning of NMDA application, but has increasing effect with increasing stimulation (Chen et al 1998). It may be that the best drugs are ones that have these moderate effects on initial currents and then become stronger as the pathological activity continues.

Noebels: Relevant to that, Ken Courtney was the first to describe this frequency-dependent reduction of phenytoin on the action potential. Later, Bob MacDonald showed that this might even be a general property of many effective anticonvulsants. I thought I understood from your presentation that you thought this was an artefact of the assay.

Segal: MacDonald's 'sustained repetitive firing' model shows a use-dependent effect of phenytoin on the Na^+ channel (McLean & MacDonald 1986) as well as a use-dependent effect of memantine (McLean 1987, Netzer & Bigalke 1990). The memantine effect turns out to be on the NMDA receptor (Chen & Lipton 1997) and not on the Na^+ channel (Lundquist 1999). I do not think that the effects uncovered by the sustained repetitive firing model are artefacts. The point I am making is that one cannot conclude that drugs that reduce sustained repetitive firing are drugs that have effects on Na^+ channels rather than on glutamate receptors.

Noebels: Could you describe the model?

Segal: The way the sustained repetitive firing model operates is that they put an electrode into a cell, depolarize the cell and it fires a lot. This is done with and without an anticonvulsant. By the fact that they are depolarizing the cell by current injection, they are getting a depolarization that in many senses is similar to the ones that we are studying as these plateau depolarizations. In contrast, we are looking at the action potentials at resting potential, and it seems that there is virtually no effect of phenytoin, as you would expect from a situation where you are dealing with an action potential that has a huge safety factor. When we study plateau depolarizations, there is no safety factor for the plateau because there is a close balance between Na^+ and K^+ currents. Any slight change is going to push the voltage one way or another.

Bean: When you see spontaneous activity after putting in the synaptic blockers, do you generally see this long-lasting plateau, even after the cell stops spiking? The plateau was clearly present in some recordings. In other cases, the cell spiked, and once it stopped spiking the voltage came down again. The plateaus that outlast the spiking seem really interesting. What I wonder is, where are the K^+ channels in these cells, and are these important in regulating that activity? Intuitively, I would think that the size of the K^+ currents in the hippocampal neurons is huge. This also raises the issue of the role of the K^+ channels in epileptic activity *in vivo*. How often do you see that plateau outlasting the spiking?

Segal: The plateau outlasted the spiking about half the time. This impressed us very much for the reason you mentioned—this suggested an equal balance between Na^+ currents and K^+ currents that we knew could be large. This is what led us to consider the possibility that persistent Na^+ currents were large, something that seemed hard to believe 10 years ago.

Bean: In most of the experiments you seemed to be using zero Ca^{2+} and high Mg^{2+} to completely block transmission. Is it the same if you just put on APV and CNQX to block transmission but leave Ca^{2+} at normal levels?

Segal: Presumably by blocking the Ca^{2+} current we are also blocking the Ca^{2+}-modulated K^+, which is probably one of the major inhibitory currents that would be occurring *in vivo*. But, on the other hand, *in vivo* you would also be having another excitatory stimulus from glutamate excitation. In our model we are knocking out one of the endogenous inhibitory actions, as well as knocking out the neurotransmitter excitation. It is probably balancing out, and we are getting some reflection of the *in vivo* voltages that would occur in such an event, but by having knocked out one excitatory and one inhibitory current we are looking at it in a simplified situation.

Bean: What happens if you put on APV and CNQX in normal Ca^{2+} and Mg^{2+}?

Segal: I don't remember any plateau depolarizations under these conditions, but we only did a few experiments of this sort before switching to using the synaptic blocking solution with zero Ca^{2+}.

Waxman: What ends the plateau?

Segal: You can end it by injecting current, but in our experiments typically we would allow it to end spontaneously, usually in several seconds.

Waxman: Do you have a sense of the events underlying this?

Segal: One would expect that the Na^+ current would decrease over time. One can't blame this on Ca^{2+} accumulation. There might be other mechanisms such as peptide secretion.

Catterall: I wanted to follow up the question that Jeff Noebels asked, which is what is the hallmark of an anticonvulsant drug as opposed to some other kind of Na^+-blocking drug? It seemed to me from our own work and reading the literature that it is preferential inactivated state block or voltage-dependent block, which is common for phenytoin, carbamazepene and lamotrigine. For example, anti-arrhythmic drugs mostly don't behave that way: they are frequency-dependent blockers. In your paradigm, do you get the same selective block of the late channel openings with carbamazepine and lamotrigine?

Segal: We tried lamotrigine, but our experiments were complicated by a DMSO solvent artefact that made it hard to get a good baseline (Douglas 1996). At some point we have to go back and do this with the newer water-soluble form of lamotrigine. We haven't tried carbamazepine. These are incredibly tedious experiments. Holding a patch over a period of 100 min is not the sort of thing you can do every day, and each experiment takes a long time to analyse.

Catterall: You could probably do this with macroscopic currents. Knowing the answer for phenytoin, you could see what happens with macroscopic currents.

Segal: We did our studies using single channel recording in part because we worried that any persistent inward current that we would see by macroscopic current recording could be dismissed as an artefact of poor clamp control of the neuron. However, Chao & Alzheimer (1995) did such a study of macroscopic currents while we were doing our single-channel studies and they found a similar result to ours. But it is hard to know how much of their quantitative result was due to direct effects on Na^+ channels or better voltage control after blockade of Na^+ channels. There are real concerns about voltage control because of the demonstration of regenerative Na^+ channel potentials in dendrites (Stuart & Sakmann 1994).

Ptacek: I presume the set of 1000 genes involved in causing epilepsy includes metabolic disorders. I'm not sure that it matters from a therapeutic point of view, but in terms of getting at the molecular mechanisms, there are two fundamentally different types of epilepsy. First, there are the asymptomatic ones where everything seems to be normal except that the membrane excitability is teetering and you get pushed over the edge. These epilepsies are much more likely to be due to variations in ion channels. However, if you fill cells up with sphingomyelin or other compounds that can't be processed normally because of an enzyme deficiency, or there is head trauma, developmental disruption or a brain tumour, I can imagine how these could give epilepsy, but those kinds of problems seem fundamentally different.

Segal: Since epilepsy is a disease of pathological excitation of neurons it is useful to consider the six final common pathways for making a neuron more excitable: changes in endogenous excitatory currents, endogenous inhibitory currents, neurotransmitter excitatory currents, neurotransmitter inhibitory currents, gap junctions and ion pumps. Changes in one of these six final common pathways could produce a simple form of epilepsy without other effects on neurons. However, a simple change such as an increase in an ionic current could also have more complicated effects if this produces changes in cell migration or cell survival. So the boundary between simple final common pathway, epilepsy, and complicated epilepsy may turn out to be very fuzzy. Even causes of epilepsy such as energy metabolism defects that clearly are not acting directly on the six final common pathways may turn out to have simple effects mediated by their effects on pumps and channels. I agree that there will be some distinction between mutations acting directly on the six final common pathways and those that do not, but the distinctions are not likely to be very crisp.

References

Baker MD, Bostock H 1998 Inactivation of macroscopic late Na^+ current and characteristics of unitary late Na^+ currents in sensory neurons. J Neurophysiol 80:2538–2549

Baker MD, Bostock H 1999 The pH dependence of late sodium current in large sensory neurones. Neuroscience 92:1119–1130

Chao TI, Alzheimer C 1995 Effects of phenytoin on the persistent Na^+ current of mammalian CNS neurons. NeuroReport 6:1778–1780

Chen HS, Lipton SA 1997 Mechanism of memantine block of NMDA-activated channels in rat retinal ganglion cells: uncompetitive antagonism. J Physiol 499:27–46

Chen, HS, Wang YF, Rayudu PV et al 1998 Neuroprotective concentrations of the N-methyl-D-aspartate open-channel blocker memantine are effective without cytoplasmic vacuolation following post-ischemic administration and do not block maze learning or long-term potentiation. Neuroscience 86:1121–1132

Douglas AF 1996 The effects of the anticonvulsant lamotrigine on the sodium channel activity of rat hippocampal neurons in microculture. Undergraduate Thesis, Harvard University, Cambridge, MA, USA

Lundquist ER 1999 Memantine as a possible anti-epileptic drug: examining its mechanism of action using models that test sodium current. Undergraduate Thesis, Harvard University, Cambridge, MA, USA

McLean MJ 1987 In vitro electrophysiological evidence predicting anticonvulsant efficacy of memantine and flunarizine. Pol J Pharmacol Pharm 39:513–525

McLean MJ, MacDonald RL 1986 Carbamazepine and 10,11-epoxycarbamazepine produce use- and voltage-dependent limitations of rapidly firing action potentials of mouse central neurons in cell culture. J Pharm Exp Ther 238:727–738

Moorman JR, Kirsch GE, VanDongen AMJ, Joho RH, Brown AM 1990 Fast and slow gating of sodium channels encoded by a single mRNA. Neuron 4:243–252

Netzer R, Bigalke H 1990 Memantine reduces repetitive action potential firing in spinal cord nerve cell cultures. Eur J Pharmacol 186:149–155

Rogawski MA, Porter RJ 1990 Antiepileptic drugs: pharmacological mechanisms and clinical efficacy with consideration of promising developmental stage compounds. Pharmacol Rev 42:223–286

Stuart GJ, Sakmann B 1994 Active propagation of somatic action potentials into neocortical pyramidal cell dendrites. Nature 367:69–72

Wallace RH, Wang DW, Singh R et al 1998 Febrile seizures and generalized epilepsy associated with a mutation in the Na^+-channel $\beta 1$ subunit gene SCN1B. Nat Genet 19:366–370

Weiss RE, Roberts WM, Stühmer W, Almers W 1986 Mobility of voltage-dependent ion channels and lectin receptors in the sarcolemma of frog skeletal muscle. J Gen Physiol 87:955–983

Therapeutic concentrations of local anaesthetics unveil the potential role of sodium channels in neuropathic pain

Gary R. Strichartz, Zhongren Zhou, Catherine Sinnott and Alla Khodorova

Pain Research Center, Department of Anesthesiology, Perioperative and Pain Medicine, Brigham and Women's Hospital, Harvard Medical School, 75 Francis Street, Boston, MA 02115-6110, USA

Abstract. Neuropathic pain is frequently associated with hyperexcitability of primary afferents, characterized by spontaneous impulses and repetitive firing. Electrophysiology and molecular biology reveal changes in dorsal root ganglion Na^+ channels under conditions of neuropathic pain, but the manner by which these changes alter the physiology of sensory afferents remains unknown. Equally mysterious is the mechanism by which i.v. local anaesthetic-like Na^+ channel blockers suppress neuropathic pain behaviour at concentrations well below those reported for channel inhibition. We have compared the anti-allodynic actions of i.v. lidocaine (L) and stereoisomers of mexiletine (*R*-M, *S*-M), in rats after spinal nerve ligation, with their ability: (1) to inhibit fast, tetrodotoxin-sensitive neuronal Na^+ currents, elicited by brief (1 ms) pulses, at 10 Hz, from 'resting' potentials (−80, −60 mV) and (2) to suppress the seconds long plateau and the repetitive firing produced in axons by slowing of Na^+ channel inactivation (e.g. using scorpion α-toxins). Both L and *R*-M at 5–10 μM relieved allodynia; *S*-M was ineffective. Na^+ currents also were inhibited by M, with affinities that were increased by both repetitive 'firing' ($K_{R,S}$ = 5 μM) and depolarization of the 'resting' membrane (K_R = 15 μM; K_S = 30 μM). Stereopotency ratios depended on the manner in which different states of the channel were inducted. Both L and M shortened the action potential's 'plateau' in α-toxin treated axons, without reducing the spike, and suppressed repetitive firing with IC_{50}s = 5 μM, and no stereoselectivity. These findings together demonstrate that Na^+ channel blockers, at 'therapeutic' concentrations, can inhibit neuronal hyperexcitability.

2002 Sodium channels and neuronal hyperexcitability. Wiley, Chichester (Novartis Foundation Symposium 241) p 189–205

Traditional local anaesthetics (lidocaine), antiarrhythmics (mexiletine) and anticonvulsants (phenytoin), antipsychotics (carbamazepine) and anti-depressants (amitriptyline) have all been used clinically, with substantial interpatient variability, to treat neuropathic pain (Ackerman et al 1991, Boas et al 1982, Chabal et al 1992). One common action of all of these diverse drugs is their ability to inhibit voltage-gated Na^+ channels (Tanelian & Victory 1995, Mao & Chen 2000). Many amphipathic amines, with a modicum of structural similarity and, at best, modest specificity, are nevertheless able to inhibit these channels, whose activity is essential for almost all membrane excitability. The question remains, however, as to whether this inhibitory action is the one most salient for neuropathic pain relief, since all of these drugs are known to have several other direct effects in the peripheral and central nervous systems.

The pathophysiological mechanism(s) for neuropathic pain is not known. Investigators have regularly detected abnormal, additional impulse activity in injured, as well as in non-injured peripheral nerves from animals with induced neuropathic pain (Kajander & Bennet 1992, Chabal et al 1989, Devor et al 1992, Study & Kral 1996, Ali et al 1999). Such 'hyperexcitability' has often been ascribed to a change in voltage-gated Na^+ channels (Matzner & Devor 1994), although it is unclear whether the salient change is an overall increase in channel density, a spatial redistribution of normal channels (Novakovic et al 1998), a shift in the expression profile of channel isoforms with different gating kinetics (Cummins et al 1999), or changes in other currents, for example K^+, that also modulate excitability.

'Hyperexcitability' in nerve injury models manifests as a reduction in electrical stimulation threshold, often to the point of generating (apparently) spontaneous impulses, and the appearance of high frequency bursts of impulses (Chabal et al 1992, Devor et al 1992, Kajander & Bennet 1992, Study & Kral 1996) which, if present in normal nociceptors, would convey a pain-like percept to the CNS. The Na^+ channel called PN3 (or SNS), detected primarily in small diameter sensory neurons (Novakovic et al 1998, Rabert et al 1998), appears to play a critical role in neuropathic pain of several aetiologies, since prevention of its expression by gene knockout (Akopian et al 1999) or by antisense oligonucleotides delivered into the spinal CSF (Porreca et al 1999) can reverse or prevent the behavioural signs of neuropathic pain. Na^+ currents through PN3 channels are longer lasting than those of other 'fast inactivating' channels (Elliott & Elliott 1993, Rush et al 1998) and can thereby support a long, post-stimulus depolarization with a period of sustained impulse firing (Parri & Crunelli 1998, Kapoor et al 1997). In opposition to a critical role of PN3 upregulation are voltage-clamp and molecular biological assays that indicate a reduction in PN3-mediated Na^+ current and messenger RNA after certain types of pain-producing procedures and suggest that a different Na^+ channel, NaN, having even more prolonged currents, may be increased (Cummins & Waxman 1997, Cummins et al 1999, Tate et al 1998). Furthermore, the reported drop in K^+ current density will also favour prolonged depolarization as well as both spontaneous and induced repetitive firing.

In this paper we report findings from neurobehavioural and *in vitro* electrophysiological investigations directed to determine if potential Na^+ channel blockers, at plasma concentrations known to be therapeutic, are indeed capable of significant Na^+ channel blockade. Using mechanical allodynia induced by spinal nerve ligation in the rat as the sign of neuropathic pain (Kim & Chung 1992), we have assessed the ability of controlled, intravenous concentrations of lidocaine and of the stereoisomers of mexiletine to restore normal force thresholds for withdrawal of the operated hindlimb. In the electrophysiological studies we have simulated nerve discharges involving non-inactivating Na^+ channels by pharmacological modification of channel gating with peptide 'α toxins' from scorpions or sea anemone, and have examined the ability of lidocaine and mexiletine to suppress the resulting abnormal impulse behaviour. Lastly, with voltage-clamp of cells having rapidly inactivating Na^+ currents, we have investigated the changes in blocking potency of these drugs due to single prolonged depolarizations or to repetitive, brief, impulse-like depolarizations.

Results and discussion

Allodynia, induced by tight ligation of the spinal nerve distal to the dorsal root ganglion (DRG), is relieved by intravenous lidocaine (L) or the *R*-isomer of mexiletine (*R*-M) (Fig. 1). Steady-state plasma drug concentrations, achieved by a computer-controlled infusion pump, define the concentration-dependence for such relief. At about 1.0 μg/ml significant alleviation of allodynia is detected for both L (Chaplan et al 1995, Sinnott et al 1999) and *R*-M (Sinnott et al 2000); the *S*-M enantiomer is ineffective (Sinnott et al 2000). At 2–4 μg/ml L and 1.2–1.4 μg/ml *R*-M, allodynic thresholds reach pre-operative levels in many animals, with average values of *c*. 60% towards complete relief. Greater relief is not attained at higher concentrations of L or *R*-M, the latter limited by toxicity (pre-convulsive twitches) at 2.0 μg/ml, as is the therapeutically ineffective *S*-M.

Notably, substantial recovery from allodynia is still present days to weeks after effective L infusions (Chaplan et al 1995, Sinnott et al 1999) whilst the peri-infusional relief by *R*-M is not followed by any residual recovery (Sinnott et al 2000), implying that separate mechanisms may underlie the acute and the persistent therapeutic phases.

Some neuropathic pain may result from spontaneous or abnormally repetitive impulses in peripheral neurons. The slowly-inactivating or even more prolonged inward currents of PN3- and NaN-type Na^+ channels, respectively, have the theoretical ability to support long-lasting depolarizations that may be accompanied by repetitive impulse activity (Kapoor et al 1997). Modest, slow depolarizations near the resting potential may stimulate impulse initiation through activation of NaN channels whose opening occurs at potentials near threshold (Kral et al

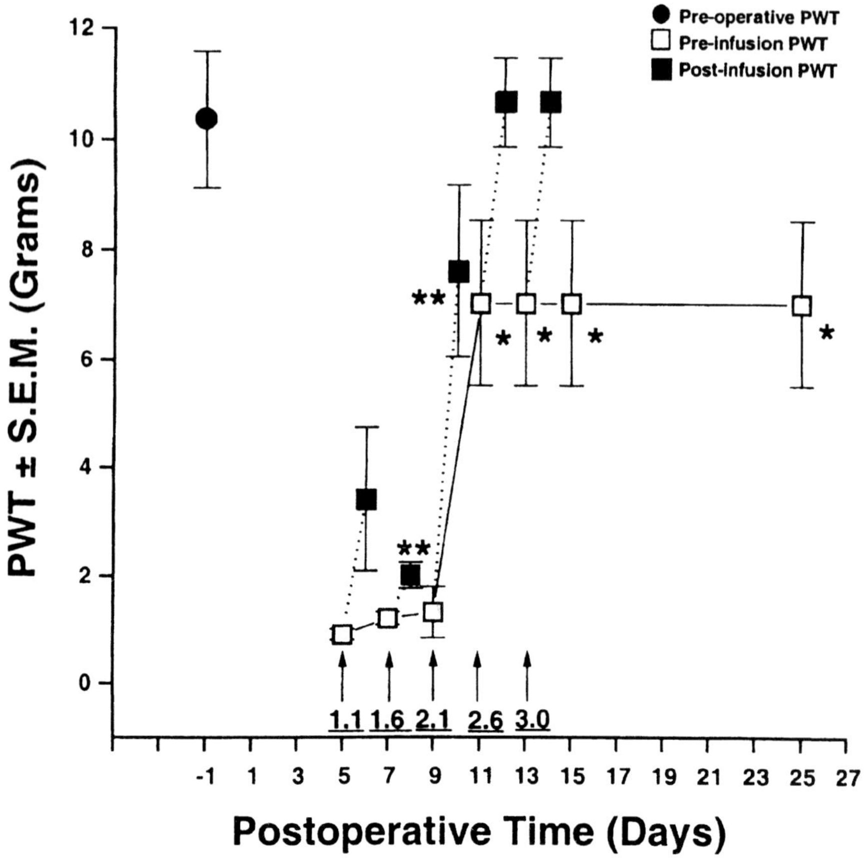

FIG. 1. The effects of successive intravenous lidocaine infusions at increasing concentrations on ipsilateral allodynia in rats ($n = 8$). Beginning on postoperative day 5, five infusions were administered at 48 h intervals which correspond to plasma concentrations of 1.11(d 5), 1.60 (d 7), 2.07 (d 9), 2.56 (d 11) and 3.03 (d 13) μg/ml. Open squares show mean paw withdrawal thresholds 5 min before administering the corresponding infusion. Filled squares show the mean paw withdrawal threshold value at the end of the infusion: *, pre-infusion threshold values on days 11, 13, 15, and 17 were significantly different from pre-infusion thresholds on day 5 ($P < 0.02$); **, post-infusion threshold values were significantly different from pre-infusion threshold values on this day ($P < 0.02$). (From Sinnott et al 1999, with permission.)

1999), much more negative than those of other Na^+ channels (Cummins et al 1999).

Normal, fast-inactivating Na^+ channels, as are found in large myelinated peripheral axons, slow their inactivation closing when bound by small peptide 'α toxins' (site III toxins) purified from scorpions or sea anemones (Ulbricht & Schmidtmayer 1981, Strichartz et al 1982). The usually brief spikes of action potentials in these fibres then manifest a subsequent 'plateau' depolarization accompanied by prolonged after-spikes (Fig. 2), with amplitude and duration dependant on the type and concentration of α toxin and on the contribution of various K^+ channels (Khodorova et al 2001). We surmise that such pharmacological manipulation simulates a pathophysiological phenotype resulting from the over-expression or maldistribution of PN3 or NaN channels.

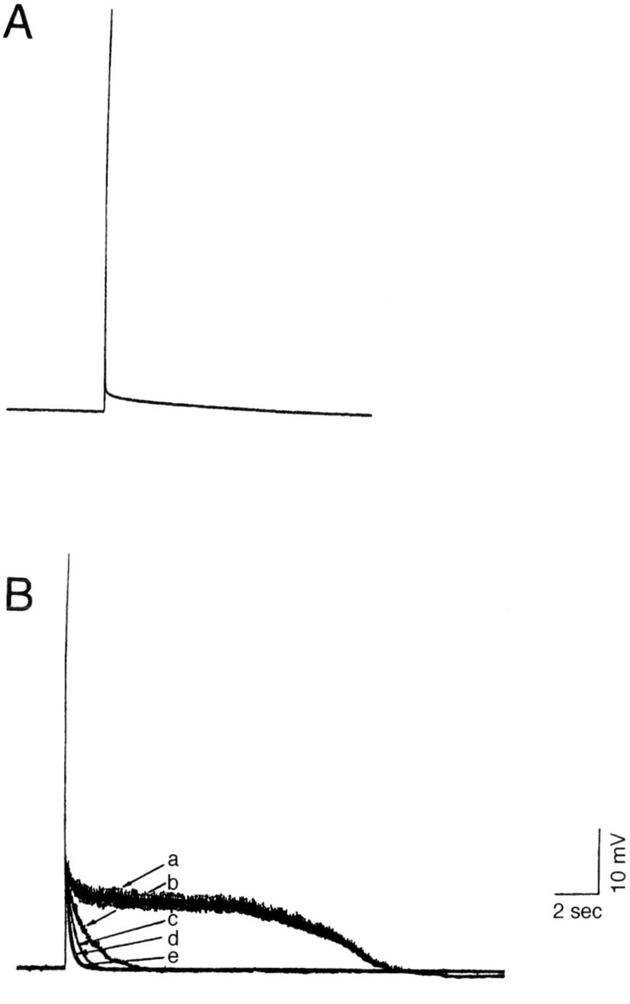

FIG. 2. Action potential recordings with long time bases show the characteristics of the plateau, which is minor in TEA·Ringer's (A), but long and 'noisy' in 0.25 μM ATXII (B), trace 'a'. Panel B also shows the shortening of the plateau by increasing concentrations of lidocaine: (a) 0, (b) 5 μM, (c) 10 μM, (d) 20 μM, (e) 30 μM. (From Khodorova et al 2001.)

Lidocaine and mexiletine both suppress the action potential features induced by α-toxins (Khodorova et al 2001). At 5–20 μM, L accelerates the decay of the plateau (Fig. 2), reduces its area by >80%, and halves the amplitudes of the high-frequency, post-spike oscillatory activity (Fig. 3). By comparison, inhibition of directly stimulated propagating spikes in peripheral fibres requires 200–800 μM L (Huang et al 1997).

Mexiletine also suppresses the α toxin-modified action potentials, with IC_{50}s equal to 1 mM for the peak and 4–6 μM for the area-under-the-plateau, both indistinguishably different between enantiomers. Half-inhibition of the first oscillation after the peak occurred at 50–60 μM *S*-M and 25–30 μM *R*-M, whilst the second oscillation's amplitude was halved by *c.* 40 μM *S*-M and 20 μM *R*-M.

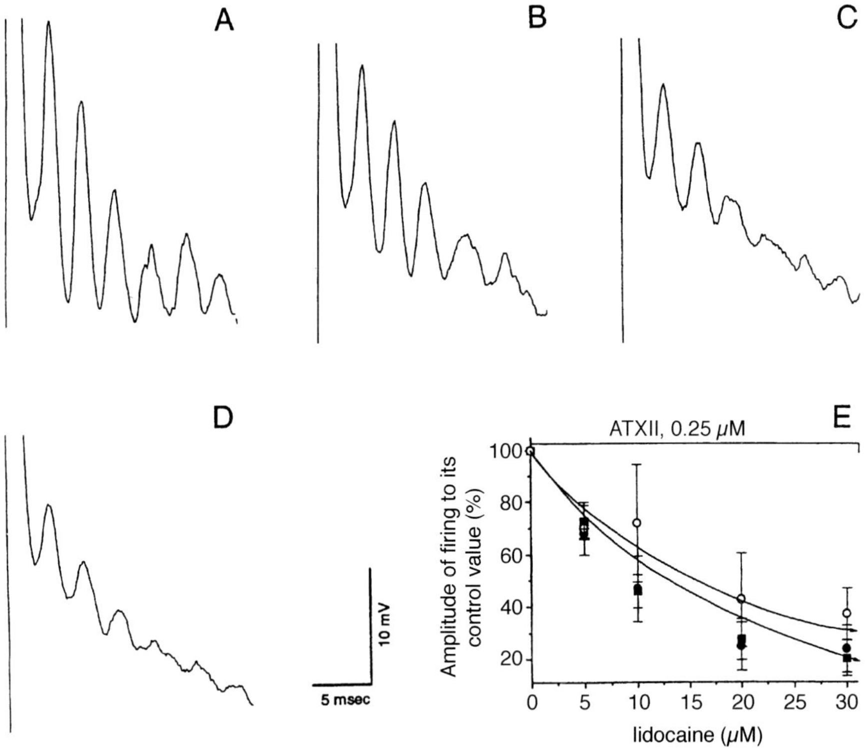

FIG. 3. Oscillations of the CAP immediately following the initial spike (truncated) in nerves exposed to ATXII (0.25 μM) plus lidocaine at concentrations of (A) 0, (B) 5 μM, (C) 10 μM or (D) 20 μM. Panel E shows the lidocaine concentration-dependent reduction of the oscillations for the first (●) and fifth (○) wave after the peak and for the average peak-to-peak noise measured from 0.05 to 0.25 s after the stimulus (■). Mean ± SEM (*n* = 3). (from Khodorova et al 2001, with permission)

These results imply that use-dependent inhibition occurs, with modest stereo-selectivity (*c*. 2), for M's actions on oscillations generated by non-inactivating Na^+ channels.

What factors account for the 100-fold difference between the concentrations required to block 'normal' impulses and those effective on plateau depolarizations and their associated 'abnormal' after-discharges? One answer, but certainly not the only one, incorporates the well-known state-dependent affinities of local anaesthetics for Na^+ channels. Lidocaine, for example, has its lowest affinity for resting, closed channels, binds with higher affinity, and relatively rapidly, to open channels and also with higher affinity, but more slowly, to inactivated, closed channels (Hille 1977, Chernoff 1990). (There is also direct evidence for enhanced binding to an intermediate state populated during the activation sequence linking resting to open channels.) The potency of these agents can be modulated by patterns of membrane potential that 'gate' the channels into different states. Theoretically, both small, slow depolarizations, such as the α-toxin induced plateau and the ramp depolarization that precedes spontaneous spikes recorded in neuropathic neurons, and rapid, brief depolarizations that occur during repetitive firing, will potentiate these drug's inhibition, through increased drug binding to inactivated channels and open channels, respectively (Chernoff & Strichartz 1989). In the following we show that, in practice, applications of these patterns of membrane potential do provide sufficient potentiation to account for an anti-allodynic effect at therapeutic concentrations.

Using voltage-clamp methods we have measured the concentration-dependence for *R*-M's and *S*-M's inhibition of peak Na^+ currents, in GH3 cells (Fig. 4A). Concentration versus inhibition curves were fit by binding equations that assumed one bound M molecule blocked one channel (Fig. 4B,C). The apparent K_I values for this inhibition fall as the 0.1s 'prepulse' membrane potential, which just precedes the current testing pulse, is made less negative. This procedure mimics the circumstance of a long plateau depolarization or, less accurately, a slow ramp depolarization before a spike. The potency of *S*-M is increased 1.5-fold by a prepulse to -60 mV, the normal resting potential of sensory neuron cell bodies (Study & Kral 1996), compared to the potency assayed at the oft-employed hyperpolarized value of -100 mV (Fig. 4B). *R*-M's potency is also enhanced by depolarizing prepulses, but by sixfold (Fig. 4C), a much larger change than *S*-M's. The fourfold increase in stereoselectivity induced by depolarizing prepulses can arise from two sources: (1) The fast inactivated state may have a higher equilibrium affinity for *R*-M or, (2) *S*-M may associate more slowly with this state, such that its binding has not equilibrated during the prepulse, whereas *R*-M's has.

The kinetic explanation for this difference was approached by holding the membrane at a constant potential before applying a test pulse. In this situation

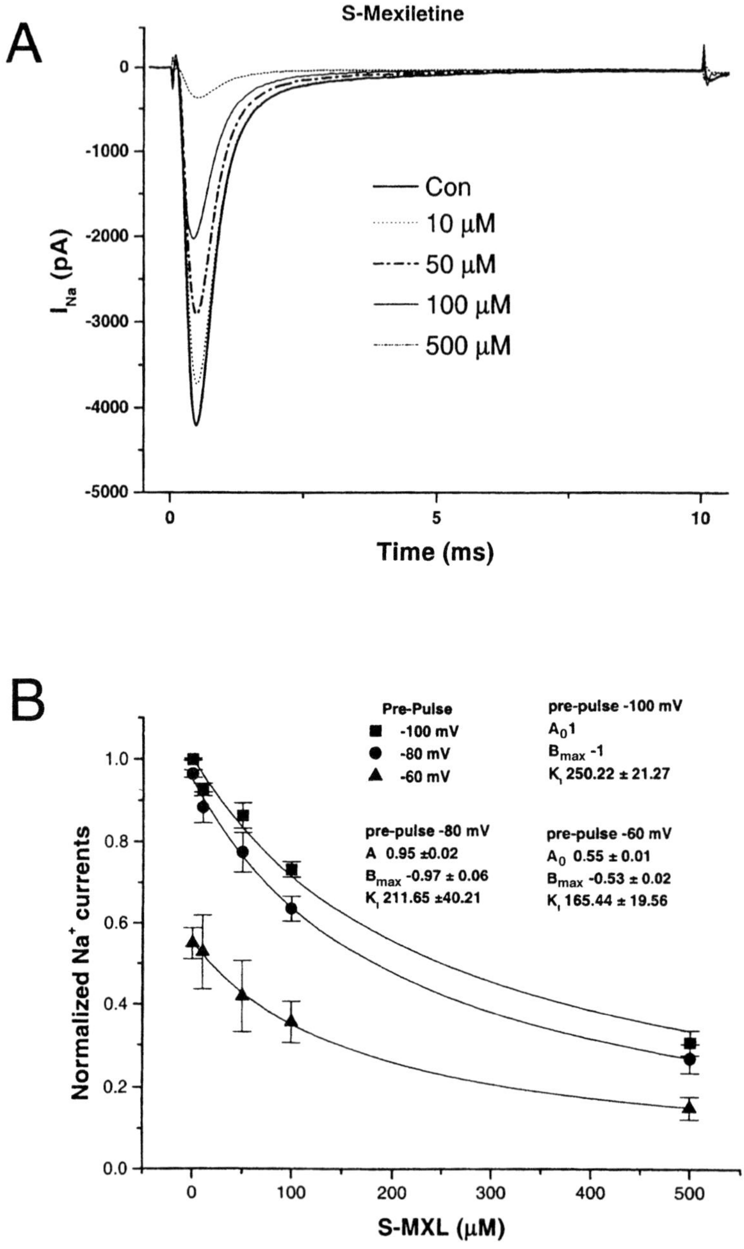
A
S-Mexiletine
I_Na (pA)
0
-1000
-2000
-3000
-4000
-5000
Con
10 μM
50 μM
100 μM
500 μM
0
5
10
Time (ms)
B
Pre-Pulse
-100 mV
-80 mV
-60 mV
pre-pulse -100 mV
A0 1
Bmax -1
Ki 250.22 ± 21.27
pre-pulse -80 mV
A 0.95 ±0.02
Bmax -0.97 ± 0.06
Ki 211.65 ±40.21
pre-pulse -60 mV
A0 0.55 ± 0.01
Bmax -0.53 ± 0.02
Ki 165.44 ± 19.56
Normalized Na+ currents
1.0
0.8
0.6
0.4
0.2
0.0
0
100
200
300
400
500
S-MXL (μM)

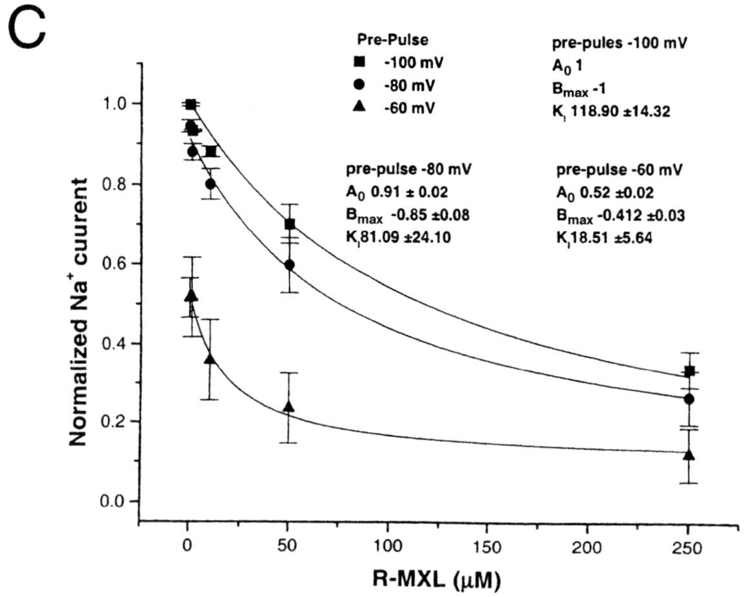

FIG. 4. Inhibition of fast Na^+ current in GH-3 cells by steroisomers of mexiletine. (A) Currents activated by 10 ms depolarizations to 0 mV from a holding potential of −80 mV are progressively diminished by increasing concentrations of *S*-M. Con is control value with no drug present. (B) The *S*-M concentration dependence of peak current inhibition is increased by depolarizing prepulses that just precede the test pulse used to measure currents. Prepulse depolarization without drug leads to channel inactivation, as represented by the availability coefficient, A_0, listed on the figure. Lines are fits of Langmuir binding isotherms to the data using Origin® software. (C) The *R*-M concentration dependence of peak current inhibition at different prepulse potentials. Details as in (B).

the apparent K_I for *R*-M changed only a small amount compared to its value for the same potential during a prepulse, for example, 60 μM for constant depolarization versus 80 μM for prepulse, both to −80 mV, or 14 μM for constant depolarization versus 20 μM for prepulse, both to −60 mV. In contrast, the K_I for *S*-M decreased dramatically for the constant depolarization: 130 μM for constant depolarization versus 210 μM for a prepulse, both to −80 mV, and 30 μM for constant depolarization versus 130 μM for a prepulse, both to −60 mV. The overall effect of a constant change in the 'resting potential' of the cell, from −100 to −60 mV, is an eightfold potentiation of *R*-M and *S*-M alike, maintaining a stereoselectivity of 2.

Brief depolarizations (1 ms duration to +10 mV) applied by voltage-clamp every 0.1 s (10 Hz) mimic the repetitive firing of impulses in neuropathic discharges. The resulting phasic, or 'use-dependent' inhibition for *R*-M (Fig. 5A), when assessed with a resting potential of $-80\,mV$, is described by an apparent inhibitory dissociation constant of $15\,\mu M$ (Fig. 5B), corresponding to a potentiation of fivefold over the value for infrequently applied pulses (Fig. 4C). A smaller potentiation (1.5-fold), describes phasic block by *S*-M over tonic inhibition from $-80\,mV$. The effect of repetitive depolarization from $-80\,mV$ has a stereopotency ratio of 4 (Fig. 5B).

When constant resting depolarization is combined with phasic stimulation (1 ms, 10 Hz) the effects on potency are further enhanced. For resting potentials of $-60\,mV$, *S*-mexiletine's channel blocking K_I becomes even lower, $5\,\mu M$, as does *R*-M's (Fig. 5C), corresponding to phasic potentiations of threefold and about sixfold, respectively, over that for an unstimulated membrane.

Altogether these data demonstrate a complex and dynamic state-dependence for inhibition of Na^+ channels by mexiletine, with the single generalization that briefer depolarizations maximize stereoselectivity but that the conditions that produce the greatest apparent affinity result in no stereoselectivity.

Of course, these are inhibitions of fast Na^+ currents in cloned, central neuroendocrine cells, not of currents in normal, not to mention 'diseased', sensory neurons. Previous reports have claimed that SNS and NaN channels are more susceptible to local anaesthetics than other, fast inactivating Na^+ channels (Roy & Narahashi 1992, Gold et al 1998, Rush & Elliott 1997, Scholz et al 1998), but an apparent potency difference can result from different distributions among various states (e.g. inactivated versus resting) rather than from intrinsic differences in affinity (Wright et al 1997). The present assays were conducted at 20 °C, not 37 °C, and temperature increases are known to reduce local anaesthetic affinity (Strichartz & Zimmermann 1983) and to shorten the channel open time, both of which lower the apparent potency for channel blockade.

Qualifications aside, the inhibition of current observed under voltage-clamp is adequate to account for the suppression of abnormal repetitive impulses and prolonged plateaus that model a neuropathic pain phenotype in α toxin-treated nerve and, most germane, these inhibitions are consonant with the plasma concentrations of drugs that relieve neuropathic pain (Abram & Yaksh 1994, Scholz et al 1998, Tanelian & MacIver 1991). Na^+ channel blockers may provide pain relief, therefore, through blockade of Na^+ channels.

Acknowledgements

The research presented here was supported by NIH grant GM35647 and by a research grant from the Celgene Corporation. Thanks to Ms. Ellen Jacobson and Ms. Belinda Brooks for secretarial services.

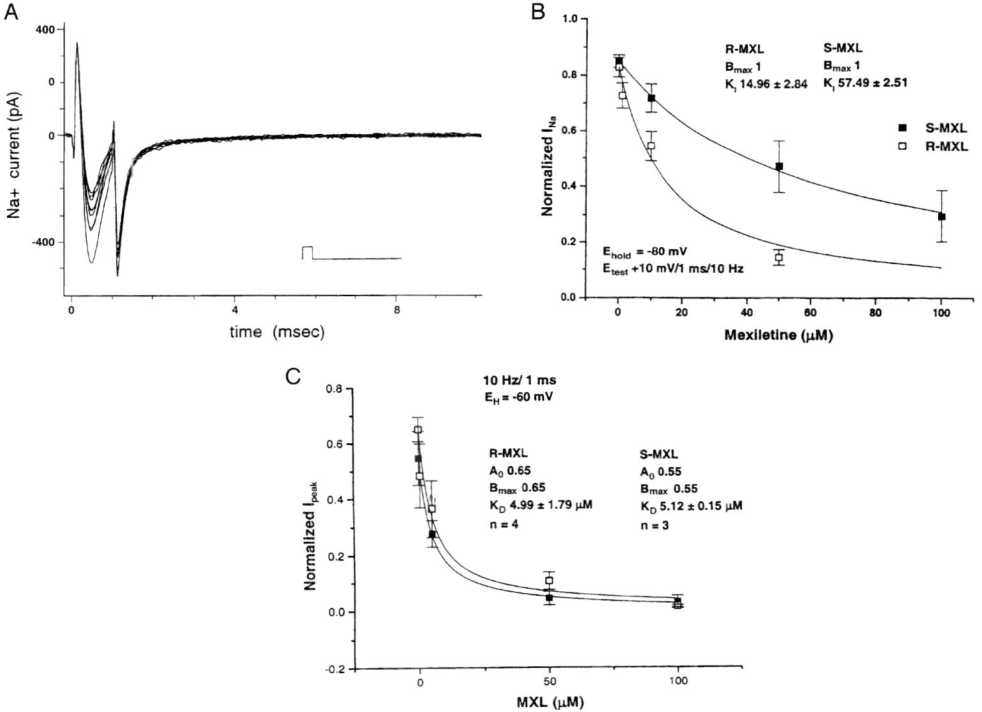

FIG. 5. Use-dependent inhibition of fast Na^+ current in GH-3 cells. (A) Tonically inhibited current (measured at 0.1 Hz frequency) is further decreased, to a new steady-state, by a train of 20, 1 s long depolarizations to 0 mV applied at 10 Hz. (B) Mexiletine's concentration-dependent role in steady-state use-dependent inhibition is shown by binding curves with inhibitory dissociation constants (K_I) that differ fourfold between enantiomers. Cumulative, endogenous slow inactivation accounts for the 15–20% reduction of current observed at 10 Hz without drugs. (C) The steady-state inhibition by M reached at 10 Hz stimulation frequency is still further potentiated when the potential between depolarizations is held at −60 mV.

References

Abram SE, Yaksh TL 1994 Systemic lidocaine blocks nerve injury-induced hyperalgesia and nociceptor-driven spinal sensitization in the rat. Anesthesiology 80:383–391

Ackerman WE, Colclough GW, Juneja MM, Bellinger K 1991 The management of oral mexiletine and intravenous lidocaine to treat chronic painful symmetrical distal diabetic neuropathy. J Ky Med Assoc 89:500–501

Akopian AN, Souslova V, England S et al 1999 The tetrodotoxin-resistant sodium channel SNS has a specialized function in pain pathways. Nat Neurosci 2:541–548

Ali Z, Ringkamp M, Hartke TV et al 1999 Uninjured C-fiber nociceptors develop spontaneous activity and alpha-adrenergic sensitivity following L6 spinal nerve ligation in monkey. J Neurophysiol 81:455–466

Boas RA, Covino BG, Shahnarian A 1982 Analgesic responses to i.v. lignocaine. Br J Anaesth 54:501–504

Chabal C, Russell LC, Burchiel KJ 1989 The effect of intravenous lidocaine, tocainide, and mexiletine on spontaneously active fibers originating in rat sciatic neuromas. Pain 38:333–338

Chabal C, Jacobson L, Mariano A, Chaney E, Britell CW 1992 The use of oral mexiletine for the treatment of pain after peripheral nerve injury. Anesthesiology 76:513–517

Chaplan SR, Bach FW, Shafer SL, Yaksh TL 1995 Prolonged alleviation of tactile allodynia by intravenous lidocaine in neuropathic rats. Anesthesiology 83:775–785

Chernoff DM 1990 Kinetic analysis of phasic inhibition of neuronal sodium currents by lidocaine and bupivacaine. Biophys J 58:53–68

Chernoff DM, Strichartz GR 1989 Binding kinetics of local anesthetics to closed and open sodium channels during phasic inhibition: relevance to anti-arrhythmic actions. In: Hondeghem LM (ed) Molecular and cellular mechanisms of anti-arrhythmic agents. Futura, New York, p 307–335

Cummins TR, Waxman SG 1997 Downregualtion of tetrodotoxin-resistant sodium currents and upregulation of a rapidly repriming tetrodotoxin-sensitive sodium current in small spinal sensory neurons after nerve injury. J Neurosci 17:3503–3514

Cummins TR, Dib-Hajj SD, Black JA, Akopian AN, Wood JN, Waxman SG 1999 A novel persistent tetrodotoxin-resistant sodium current in SNS-null and wild-type small primary sensory neurons. J Neurosci 19:RC43

Devor M, Wall PD, Catalan N 1992 Systemic lidocaine silences ectopic neuroma and DRG discharge without blocking nerve conduction. Pain 48:261–268

Elliott AA, Elliott JR 1993 Characterization of TTX-sensitive and TTX-resistant sodium currents in small cells from adult rat dorsal root ganglia. J Physiol 463:39–56

Gold MS, Reichling DB, Hampl KF, Drasner K, Levine JD 1998 Lidocaine toxicity in primary afferent neurons from the rat. J Pharmacol Exp Ther 285:413–421

Hille B 1977 Local anesthetics: hydrophilic and hydrophobic pathways for the drug-receptor reaction. J Gen Physiol 69:497–515

Huang JH, Thalhammer JG, Raymond SA, Strichartz GR 1997 Susceptibility to lidocaine of impulses in different somatosensory afferent fibers of rat sciatic nerve. J Pharmacol Exp Ther 282:802–811

Kajander KC, Bennett GJ 1992 Onset of a painful peripheral neuropathy in rat: A partial and differential deafferentation and spontaneous discharge in Aβ and AΔ primary afferent neurons. J Neurophysiol 68:734–744

Kapoor R, Li YG, Smith KJ 1997 Slow sodium-dependent potential oscillations contribute to ectopic firing in mammalian demyelinated axons. Brain 120:647–652

Khodorova A, Konrad Meissner K, Leeson S, Strichartz GR 2001 Lidocaine selectively blocks abnormal impulses arising from noninactivating Na channels. Muscle Nerve 24:634–647

Kim SH, Chung JM 1992 An experimental model for peripheral neuropathy produced by segmental spinal nerve ligation in the rat. Pain 50:355–363

Kral MG, Xiong Z, Study RE 1999 Alteration of Na^+ currents in dorsal root ganglion neurons from rats with a painful neuropathy. Pain 81:15–24

Mao J, Chen LL 2000 Systemic lidocaine for neuropathic pain relief. Pain 87:7–17

Matzner O, Devor M 1994 Hyperexcitability at sites of nerve injury depends on voltage-gated sensitive Na^+ channels. J Neurophysiol 72:349–359

Novakovic SD, Tzoumaka E, McGivern JG et al 1998 Distribution of the tetrodotoxin-resistant sodium channel PN3 in rat sensory neurons in normal and neuropathic conditions. J Neurosci 18:2174–2187

Parri HR, Crunelli V 1998 Sodium current in rat and cat thalamocortical neurons: role of a non-inactivating component in tonic and burst firing. J Neurosci 18:854–867

Porreca F, Lai J, Bian D et al 1999 A comparison of the potential role of the tetrodotoxin-insensitive sodium channels, PN3/SNS and NaN/SNS2, in rat models of chronic pain. Proc Natl Acad Sci USA 96:7640–7644

Rabert DK, Koch BD, Klnicka M et al 1998 A tetrodotoxin-resistant voltage-gated sodium channel from human dorsal root ganglia, hPN3/SCN10A. Pain 78:107–114

Roy ML, Narahashi T 1992 Differential properties of tetrodotoxin-sensitive and tetrodotoxin-resistant sodium channels in rat dorsal root ganglion neurons. J Neurosci 12:2104–2111

Rush AM, Bräu ME, Elliott AA, Elliott JR 1998 Electrophysiological properties of sodium current subtypes in small cells from adult rat dorsal root ganglia. J Physiol 511:771–789

Rush AM, Elliott JR 1997 Phenytoin and carbemazepine: differential inhibition of sodium currents in small cells from adult rat dorsal root ganglia. Neurosci Lett 226:95–98

Scholz A, Kuboyama N, Hempelmann G, Vogel W 1998 Complex blockade of TTX-resistant Na^+ currents by lidocaine and bupivacaine reduce firing frequency in DRG neurons. J Neurophysiol 79:1746–1754

Sinnott CJ, Garfield JM, Strichartz G 1999 Differential efficacy of intravenous lidocaine in alleviating ipsilateral versus contralateral neuropathic pain in the rat. Pain 80:521–531

Sinnott CH, Garfield JM, Zeitlin A et al 2000 Enantioselective relief of neuropathic pain by systemic mexiletine in the rat. J Pain 1:128–137

Strichartz G, Hahin R, Cahalan M 1982 Pharmacological models for sodium channels producing abnormal impulse activity. In: Culp W, Ochoa J (eds) Nerves and muscles as abnormal impulse generators. Oxford University Press, Oxford p 98–129

Strichartz GR, Zimmerman M 1983 Selective conduction blockade among different fiber types in mammalian nerves by lidocaine combined with low temperature. Soc Neuroscience 13th Annual Meeting Abstracts p 675

Study RE, Kral MG 1996 Spontaneous action potential activity in isolated dorsal root ganglion neurons from rats with a painful neuropathy. Pain 65:235–242

Tanelian DL, MacIver MB 1991 Analgesic concentrations of lidocaine suppress the tonic A-delta and C fiber discharges produced by acute injury. Anesthesiology 74:934–936

Tanelian DL, Victory RA 1995 Sodium channel-blocking agents: Their use in neuropathic pain conditions. Pain Forum 4:75–80

Tate S, Benn S, Hick C, Trezise D et al 1998 Two sodium channels contribute to the TTX-R sodium current in primary sensory neurons. Nat Neurosci 1:653–655

Ulbricht W, Schmidtmayer J 1981 Modification of sodium channels in myelinated nerve by Anemonea sulcatata toxin II. J Physiol Paris 77:1103–1111

Wright SN, Wang SY, Kallen RG, Wang GK 1997 Differences in steady-state inactivation between Na channel isoforms affect local anesthetic binding affinity. Biophys J 73:779–788

DISCUSSION

Wood: Can we argue that the effective concentration of lidocaine in the *in vivo* experiments is very much higher than that you measure in the perfusate? Perhaps it concentrates in certain tissues. Have you measured the real concentration in neuropathic pain in peripheral tissues?

Strichartz: No we haven't. We are planning on measuring the concentrations in brain, spinal cord and peripheral nerve since it is easy to get radiolabelled lidocaine. We believe that we have reached a steady state, and we reach the steady state of neuropathic pain relief 5–10 min after the infusion starts. Although you might argue that it is more concentrated in the more hydrophobic cores of the tissue, you would expect the same relationship between the membrane concentration and the free concentration in equilibrium in an isolated tissue as you would *in vivo*. Although you might think of it as more concentrated per cubic centimetre of tissue, the effective activity of the drug would be the same.

Waxman: While we are talking about concentrations of various molecules as they approach the node, years ago Landon and Langley showed that nodes of Ranvier in peripheral nerve are surrounded by a cuff of what they called 'gap substance'. They subsequently presented some evidence that this acts as a cation binding lattice. There is also evidence that the gap substance can bind some anions. For those of you who do immunocytochemistry, this extracellular substance also binds many antibodies, so when you do immunocytochemistry on axons you not only see membrane fluorescence, but also artefactual or non-membrane-associated binding of antibodies to this extracellular material surrounding the node. Gary Strichartz, do you want to comment on the issue of what concentrations of TTX, for example, are seen at a node of Ranvier, compared to those that you applied to a whole nerve?

Strichartz: You are implying that if you have a local concentration of some negatively charged substance that is fixed near a binding site, then mobile charged ligands will be concentrated in that area. This is certainly true. But for chemical purposes, what you want to ask is what is the electrostatic potential within a mean free path of the binding site? If you have an electrostatic potential that is 10 Å away, which is about 1 Debye length in normal solution, then this potential has fallen substantially by the time you get to the receptor. It is really the potential at the membrane surface, in the so-called diffuse double layer, that is important. I don't know the extent to which the extranodal substance gets close enough to ion channels to effect an increase in concentration of charged ligands.

Bean: If I remember correctly, lidocaine at neutral pH is split roughly equally between the neutral and protonated form. Is this different from mexiletine? Could this be one of the elements that contributes to their difference in giving the persistent relief of pain that you see with lidocaine and not with mexiletene? Or do you think this reflects an action on a completely different target?

Strichartz: Persistent relief means that one week later you still have relief, and all the lidocaine would have cleared from the animal by then.

Bean: I was thinking that there might be different actions on the Na^+ channels while the drug is there.

Strichartz: If you look at the Na^+ currents in different tissues with lidocaine and mexiletene, there is nothing obviously different. The potencies are a little different, as are the pK_as. The octanol:buffer partition coefficients, which people usually equate with lipophilicities, are roughly the same. We think the two molecules have a totally different mechanism.

Bean: Have you any idea what the target is?

Strichartz: No.

Meisler: Are there any good data on the half-lives of the channel proteins *in vivo*?

Strichartz: I don't think so. In cultured cells it is a matter of days. If glycosylation is prevented it is hours.

Catterall: We did experiments in cultured primary neurons from rat brain. The Na^+ channels turn over with a half-life of about 48 h in pulse–chase experiments. But I don't imagine that a myelinated neuron that runs all the way down to your foot turns over its Na^+ channels in 48 h.

Strichartz: Although there is this large pool of α subunits that are present inside nodes, are they just sitting there waiting for something unusual to happen? This seems to be a great expenditure of biosynthesis for nothing.

Catterall: We also found in the experiments with the cultured neurons that there is a large intracellular pool of Na^+ channels that are incompletely assembled with β subunits. Na^+ channels must be more stable in mature nerves.

Waxman: When we looked at STX binding in Schwann cells, 50% of the binding sites were not on the membrane but in the intracellular compartment.

Bevan: You have shown the behavioural data from rats and then switched to the frog peripheral nerve, making the connection that the effects you are seeing *in vivo* are in fact mediated by effects on peripheral nerve. But I think that there is an argument that for some local anaesthetics, some of the anti-nociception effects are centrally mediated.

Strichartz: I don't think we know whether they are acting on the peripheral nerve or not. We were testing the hypothesis that abnormal activity, which has been reported from peripheral nerve in a number of different procedures, might be the substrate for local anaesthetics. But you are right: there has been a fair amount of work in spinal cord looking at a number of different synaptic properties (both peptidergic transmission and glycine/glutamate transmission) showing that low concentrations of local anaesthetic are able to alter transmission.

Spruston: One thing I noticed is that you didn't show any examples of bursting, even in the presence of the scorpion toxins. Do you ever see that? I know some people have reported action potential bursting in DRG neurons (White et al 1989).

Strichartz: It could be there. We are looking at the measured discharge from about 10^3 fibres. Although they are in synchrony right after the initial spike, they could go into a burst mode later on. If they were asynchronous at that time, you wouldn't be able to see it, so you would have to tease out a small fibre to look for this. We have looked in peripheral nerve at single units in an anaesthetized rat *in vivo*, but we have never seen a peripheral sensory fibre that showed any burst activity.

Horn: What causes those oscillations that you are seeing?

Strichartz: The α toxin concentrations we use would modify about 10% of the Na^+ channels in the nerve. We know this from our other voltage clamp studies. So 85–90% of the channels are normal. After a spike, the non-inactivating channels provide sufficient current to maintain this plateau. This is at a potential that is sufficient to activate the normal channels, which have closed after the spike and will open again. If you use a higher concentration of toxin, as I showed with 30 nM *Leiurus* toxin, you get a larger plateau, but there are no oscillations there. Our argument is that these large depolarizations have inactivated a substantial fraction of toxin-free channels, now about a third of the total channels.

Horn: This is a compound action potential, so the oscillation frequency is the same in a lot of the fibres.

Strichartz: Yes, and we are using a stimulus which selectively activates the largest myelinated fibres. We don't get slower fibres whose later pulses might interfere with the signal that we can measure.

Bean: Do you know if you get oscillations and synchronized firing in the allodynia model? Can you take nerves from those animals and see the oscillatory activity?

Strichartz: People have reported spontaneous discharges in Aβ fibres from rats treated this way. I don't remember what the frequency of firing is. I think it is between 50 and 100 Hz.

Gold: What Marshall Devor's group has reported is basically three patterns of activity. One they call tonic firers, another contains bursters with fairly large interburst intervals and the third are irregular firers. The proportions of these different firing types apparently change throughout the duration of the injury. You start out relatively early on with more of the tonic firing, and then at later dates you move to more phasic and irregular firing.

Strichartz: Aren't those data from neuromas?

Gold: No, these are recent data from the SNL model.

Bean: Is it known whether this is originating from the cell bodies or the middle of the axon, for example?

Gold: This is dorsal root recording. The assumption is that it is somewhere around the ganglia. If you cut proximal to the ganglia you eliminate the activity.

Strichartz: Marshall Devor and colleagues studied the discharges in the rat sciatic nerve after neuroma formation, where they sectioned the nerve and then put a cap over the end, so it grows back into some tangled mess (Devor et al 1992). Neuromas have their own pathophysiology, and he gave intravenous bolus injections of lidocaine, but didn't measure concentrations. He showed that the activity he could record near the ganglion was the most susceptible (i.e. it disappeared first), activity at the site of injury was next and then activity recorded in branches that were heading centrally was the least susceptible. Whether this was just because of kinetic distribution of the drug is unknown. We are planning to do similar experiments with constant infusions of lidocaine to different levels.

Gold: Is it known whether mexiletene blocks the Ca^{2+} ATPase?

Strichartz: I don't know.

References

Devor M, Wall P, Catalan N 1992 Systemic lidocaine silences ectopic neuroma and DRG discharge without blocking nerve conduction. Pain 48:261–268

White G, Lovinger DM, Weight FF 1989 Transient low-threshold Ca^{2+} current triggers burst firing through an afterdepolarizing potential in an adult mammalian neuron. Proc Natl Acad Sci USA 86:6802–6806

Molecular mechanisms of gating and drug block of sodium channels

William A. Catterall

Department of Pharmacology, University of Washington, Box 357280, Seattle, WA 98195-7280

Abstract. Voltage-gated Na^+ channels are composed of an α subunit of 260 kDa associated with β subunits of 33–36 kDa. α subunits have four homologous domains (I to IV) containing six transmembrane α helices (S1–S6). The S4 segments serve as voltage sensors and move outward to initiate activation. The S5 and S6 segments and the short membrane-associated loops between them form the pore. Fast inactivation is mediated by closure of an inactivation gate formed by the intracellular loop between domains III and IV. The 3-D structure of the inactivation gate has been determined by NMR spectroscopy, revealing the conformation of the pore-blocking IFM motif. Peptide scorpion toxins that alter gating of Na^+ channels bind to the extracellular ends of the IIS4 and IVS4 segments, trap them in either an activated or non-activated position, and thereby selectively alter channel activation or inactivation. Voltage sensor-trapping may be a general mechanism of toxin action on voltage-gated ion channels. Local anaesthetics block the pore of Na^+ channels by binding to a receptor site in segment S6 in domains III and IV. Anticonvulsants and antiarrhythmic drugs also interact with this site. A high-affinity Na^+ channel blocker has recently been developed with this site as its target. The emerging knowledge of the molecular mechanisms of Na^+ channel gating and drug block may allow development of novel therapeutics for epilepsy, cardiac arrhythmia and persistent pain syndromes.

2002 Sodium channels and neuronal hyperexcitability. Wiley, Chichester (Novartis Foundation Symposium 241) p 206–225

Na^+ channels from mammalian brain are complexes of α (260 kDa), $\beta 1$ (36 kDa) and $\beta 2$ (33 kDa) subunits (Hartshorne & Catterall 1981, Hartshorne et al 1982). The Na^+ channel α subunit was first cloned from electric eel electroplax (Noda et al 1984), but successful functional expression of the Na^+ channel required cloning cDNAs from rat brain. RNA encoding the α subunit proved sufficient for functional expression of Na^+ currents in *Xenopus* oocytes (Noda et al 1986a, 1986b, Goldin et al 1986), but the β subunits are required for normal kinetics and voltage dependence of gating (Isom et al 1992, 1995).

Application of molecular modelling to the Na^+ channel (Guy & Seetharamulu 1986) predicted six α-helical transmembrane segments (S1–S6) in each of the four

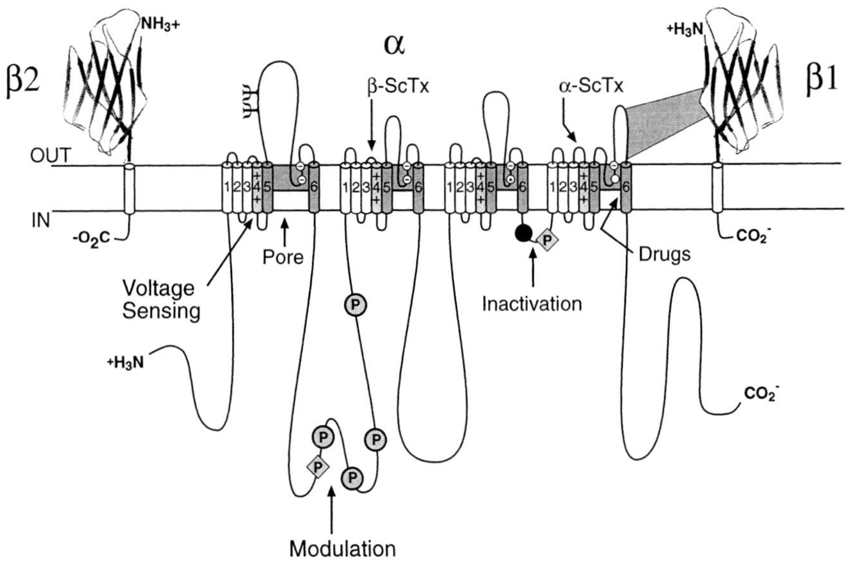

FIG. 1. Subunit structure of the voltage-gated Na^+ channels. The primary structures of the subunits of the voltage-gated ion channels are illustrated as transmembrane folding diagrams. Cylinders represent probable α helical segments. Bold lines represent the polypeptide chains of each subunit with length approximately proportional to the number of amino acid residues in the brain Na^+ channel subtypes. The extracellular domains of the $\beta 1$ and $\beta 2$ subunits are shown as immunoglobulin-like folds. Ψ, sites of probable *N*-linked glycosylation; P in shaded circles, sites of demonstrated protein phosphorylation by PKA (circles) and PKC (diamonds); shaded S5 and S6, pore-lining segments; clear circles, the outer (EEEE) and inner (DEKA) rings of amino acid residues that form the ion selectivity filter and the tetrodotoxin binding site; shaded, S4 segments, voltage sensors; dark circle, inactivation particle in the inactivation gate loop. Sites of binding of α and β scorpion toxins and a site of interaction between α and $\beta 1$ subunits are also shown.

homologous domains (I–IV) and a re-entrant loop that dipped into the transmembrane region of the protein between transmembrane segments S5 and S6 and formed the outer pore (Fig. 1). Relatively large extracellular loops were predicted in each homologous domain, connecting either the S5 or S6 transmembrane segments to the membrane-reentrant loop. Even larger intracellular loops were predicted to connect the four homologous domains, and large N-terminal and C-terminal domains were also predicted to be intracellular. Subsequent work on Na^+, Ca^{2+} and K^+ channels is consistent with the general features of this early model.

Comparison of the primary structures of the auxiliary $\beta 1$ and $\beta 2$ subunits to those of other proteins revealed a clear structural relationship to the family of proteins that contain immunoglobulin-like folds (Isom et al 1995). The extracellular domains of the $\beta 1$ and $\beta 2$ subunits are predicted to fold in a similar manner as myelin protein P_0, whose structure is known (Fig. 1; Shapiro et al 1996). The $\beta 1$ and $\beta 2$ subunits of the Na^+ channel appear to have dual functions— modulation of channel gating and cell–cell interaction. The effect of the $\beta 1$ subunit on Na^+ channel activation and inactivation is mediated by the immunoglobulin-like fold in the extracellular domain (McCormick et al 1998), which is sufficient to modulate channel gating when attached to an unrelated transmembrane segment or to a glycophospholipid anchor without the transmembrane and intracellular domains (McCormick et al 1999). Analysis of two different sets of channel chimeras points to the loop on the extracellular side of transmembrane segment IVS6 as one important point of interaction of the $\beta 1$ subunit (Fig. 1, Makita et al 1996, Qu et al 1999). Interactions with this extracellular loop serve to modulate channel activation and coupling to fast inactivation via an unknown mechanism.

The structural similarity of the β subunits to cell adhesion molecules suggests that they perform similar functions, and direct experimental support for this idea has come from recent experiments on interaction of the β subunits with extracellular proteins. Na^+ channels, and the $\beta 2$ subunit, bind to the extracellular matrix proteins tenascin C and tenascin R (Srinivasan et al 1998). Transfected cells expressing Na^+ channel subunits are repelled by surfaces coated with tenascin R, as though interaction with this extracellular protein is a repellent signal to migrate away from the interacting surface (Xiao et al 1999, Isom 2002, this volume). These interactions may guide the formation of specialized areas of high Na^+ channel density such as nodes of Ranvier and axon initial segments and may stabilize the high density of Na^+ channels in these locations.

Voltage-dependent activation and voltage-sensor trapping by β scorpion toxins

The voltage dependence of activation of the Na^+ channel and other voltage-gated ion channels derives from the outward movement of gating charges in response to changes in the membrane electric field (Armstrong 1981, Horn 2002, this volume). Approximately 12 electronic charges in the Na^+ channel protein move across the membrane electric field during activation (Hirschberg et al 1995). The novel features of the primary structure of the Na^+ channel α subunit led directly to hypotheses for the molecular basis of voltage-dependent gating (Guy & Seetharamulu 1986, Catterall 1986). The S4 transmembrane segments contain repeated motifs of a positively charged amino acid residue followed by two hydrophobic residues, potentially creating a cylindrical α helix with a spiral ribbon of positive charge around it. The negative internal transmembrane electrical field would exert a strong force on these positive charges arrayed across the plasma membrane, pulling them into the cell in a cocked position.

Depolarization of the membrane is proposed to release the S4 segments to move outward along a spiral path, initiating a conformational change that opens the pore. The proposed outward movement of the S4 segments of Na^+ channels has been directly measured using Na^+ channel mutants with cysteine residues substituted for positively charged amino acid residues (Yang & Horn 1995, Yang et al 1996, Horn 2002, this volume). Remarkably, these experiments showed that three positively charged amino acid residues in the S4 segment of domain IV become accessible outside the cell during channel gating.

Our recent experiments show that the outward gating movement of the S4 segments during channel activation is also a target for neurotoxin action. β scorpion toxins enhance Na^+ channel activation by shifting its voltage dependence to more negative membrane potentials. We analysed toxin binding and action on a family of chimeric Na^+ channels in which one of the extracellular loops of the brain Na^+ channel was replaced with its counterpart from the toxin-insensitive cardiac Na^+ channel (Cestèle et al 1998). The results show that β scorpion toxins bind to a receptor site including the S3–S4 extracellular loop near the extracellular end of the IIS4 transmembrane segment (Cestèle et al 1998). Toxin binding alone has no effect on activation, but when the channel is activated by depolarization, the bound toxin enhances activation by negatively shifting its voltage dependence (Fig. 2). Once activated, the presence of the toxin greatly slows the deactivation of the channel, creating large tail currents. This effect is thought to be mediated by trapping the activated S4 voltage sensor in its outward, activated position by binding of the S4 segment to the toxin sitting in its receptor site on the extracellular surface of the channel protein (Fig. 2; Cestèle et al 1998). A prediction of this model is that mutations that allow the IIS4 segment to move outward more easily would enhance toxin action. We found that mutants in which the positive gating charges in the IIS4 segment were neutralized by substitution of glutamine or cysteine for arginine were much more sensitive to enhanced activation by β scorpion toxins. In these mutants toxin binding alone is sufficient for voltage sensor trapping without prior activation of the channel. Voltage-sensor trapping appears to be a widespread mechanism through which channel gating is altered by polypeptide neurotoxins, including the effects of α scorpion toxins on Na^+ channel inactivation described below (Rogers et al 1996).

Structural basis for Na^+ channel inactivation

Na^+ channels inactivate within a few milliseconds of opening. Based on its sensitivity to proteases perfused inside the squid giant axon, fast inactivation was thought to be mediated by an intracellular gate that binds to the intracellular mouth of the pore (Armstrong 1981). Site-directed anti-peptide antibodies against the short, highly conserved intracellular loop connecting domains III and

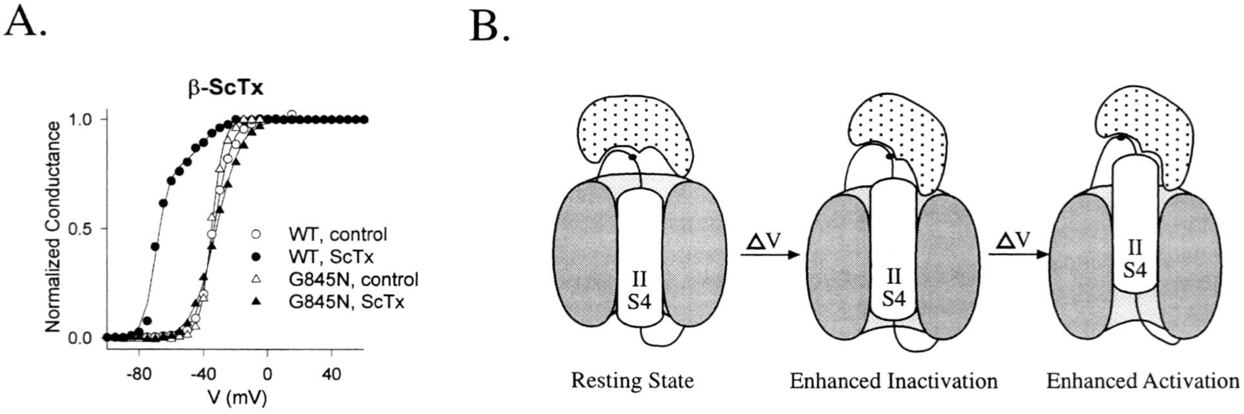

FIG. 2. Voltage sensor trapping by β scorpion toxin. (A) Effect of β scorpion toxin on activation of wild-type and mutant Na^+ channels. Na^+ currents were measured at the indicated test pulse potentials by whole-cell voltage clamp in tsA-201 cells transfected with wild-type $Na_v1.2$ channels or mutant G845N (Cestèle et al 1998). Where indicated by ScTx, the β scorpion toxin CssIV was present at a saturating concentration (Cestèle et al 1998). (B) Voltage sensor trapping model for β scorpion toxin action. Domain II and its S4 voltage sensor are illustrated at a negative resting membrane potential (e.g. −120 mV), a more depolarized resting membrane potential (e.g. −70 mV), and a fully depolarized membrane potential (e.g. +40 mV).

IV of the Na^+ channel α subunit (Fig. 1) prevent fast Na^+ channel inactivation (Vassilev et al 1989, 1988). Moreover, the accessibility of this site for antibody binding was reduced when the membrane was depolarized to induce inactivation, suggesting that the loop connecting domains III and IV forms an inactivation gate which folds into the channel structure during inactivation (Vassilev et al 1989, 1988). Cutting the loop between domains III and IV by expression of the Na^+ channel in two pieces greatly slows inactivation (Stuhmer et al 1989). Mutagenesis studies of this region revealed a hydrophobic triad of isoleucine, phenylalanine and methionine (IFM) that is critical for fast inactivation (Fig. 1, dark circle; West et al 1992), and peptides containing this motif can serve as pore blockers and can restore inactivation to Na^+ channels having a mutated inactivation gate (Eaholtz et al 1994). These results support a model in which the IFM motif serves as a tethered pore blocker that binds to a receptor in the intracellular mouth of the pore (Fig. 3A). Inactivation is impaired in proportion to the hydrophilicity of amino acid substitutions for the key phenylalanine residue (F1489), suggesting that it enters into a hydrophobic interaction with an inactivation gate receptor during inactivation (Kellenberger et al 1997). Voltage-dependent movement of the inactivation gate has been detected by measuring the accessibility of a cysteine residue substituted for F1489 (Kellenberger et al 1996). This substituted cysteine residue becomes inaccessible to reaction with sulfhydryl reagents as the inactivation gate closes.

The 3-D structure of the central portion of the inactivation gate has been determined by expression as a separate peptide and analysis by multi-dimensional NMR methods (Rohl et al 1999). These experiments reveal a rigid α helix flanked on its N-terminal side by two turns, the second of which contains the IFM motif (Fig. 3B). In this position, F1489 is poised to serve as a tethered ligand that occludes the pore. The nearby threonine (T1491), which is an important residue for inactivation (Kellenberger et al 1997), is also in a position to interact with the inactivation gate receptor in the pore. In contrast, the methionine of the IFM motif (M1490) is buried in the core of the peptide, interacting with two tyrosine residues in the α helix. This hydrophobic interaction stabilizes the fold of the peptide and forces F1489 into its exposed position. The structure of the inactivation gate peptide in solution suggests that the rigid α helix serves as a scaffold to present the IFM motif and T1491 to a receptor in the mouth of the pore as the gate closes.

Coupling of activation to inactivation

Na^+ channel inactivation derives most or all of its voltage dependence from coupling to the activation process driven by transmembrane movements of the S4 voltage sensors (Armstrong 1981). Increasingly strong evidence implicates the S4 segment in domain IV in this process. Mutations of charged amino acid

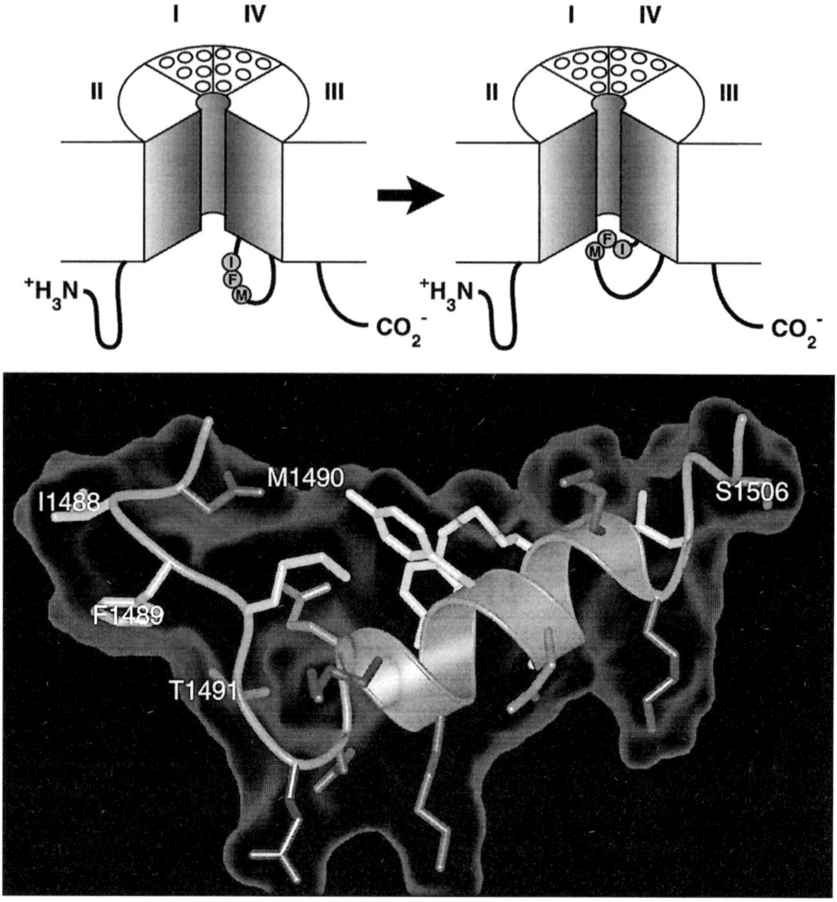

FIG. 3. Mechanism of inactivation of Na^+ channels. (A) The hinged-lid mechanism of Na^+ channel inactivation is illustrated. The intracellular loop connecting domains III and IV of the Na^+ channel is depicted as forming a hinged lid. The critical residue phenylalanine 1489 (F) is shown occluding the intracellular mouth of the pore in the Na^+ channel during the inactivation process. (B) 3D structure of the central segment of the inactivation gate as determined by multidimensional NMR (Rohl et al 1999). Isoleucine 1488, phenylalanine 1489, and methionine 1490 (IFM) are illustrated. Threonine 1491, which is important for inactivation, and serine 1506, which is a site of phosphorylation and modulation by protein kinase C, are also indicated.

residues at the extracellular end of the IVS4 segment have strong and selective effects on inactivation (Chen et al 1996). Toxins that slow coupling of activation to inactivation bind to a receptor site on the extracellular side of the IVS4 segment (Rogers et al 1996). The IIIS4 and IVS4 segments, detected by covalently incorporated fluorescent probes, are specifically immobilized in the outward position by fast inactivation, arguing that their movement is coupled to the inactivation process (Cha et al 1999). Together, these results provide strong evidence that outward movement of the S4 segment in domain IV is the signal to initiate fast inactivation of the Na^+ channel by closure of the intracellular inactivation gate

As for the IIS4 segment, the movement of the IVS4 segment is also a molecular target for neurotoxin action. α scorpion toxins and sea anemone toxins uncouple activation from inactivation. They bind to a receptor site including the S3–S4 loop at the extracellular end of the IVS4 segment (Rogers et al 1996, Sheets et al 1999). A single mutation of R1613 in this extracellular loop reduces binding affinity and thus slowing of inactivation by more than 50-fold (Fig. 4). Binding of the toxins in this position is proposed to slow inactivation by preventing the normal outward movement of the IVS4 segment (Fig. 4), evidently trapping it in a position that is permissive for activation but not for fast inactivation. Thus, scorpion venoms contain two different toxins that act by voltage sensor-trapping—the β scorpion toxins trap the IIS4 segment in an activated position (Fig. 2) and enhance activation while the α scorpion toxins trap the IVS4 segment in an inward, partially activated position that allows activation but not fast inactivation (Fig. 4). The combination of the two effects increases Na^+ channel activity inappropriately and causes paralysis. The differential effects of these two toxins also reinforce the evidence for specialization of the S4 segments in different domains—domain II for activation and domain IV for coupling of activation to inactivation.

The inner pore and local anaesthetic receptor site

Voltage clamp studies led to the conclusion that local anaesthetics enter from the intracellular side and bind in the inner pore of Na^+ channels (Strichartz et al 2002, this volume) and similar work revealed analogous intracellular block of Ca^{2+} and K^+ channels. The first indication that the S6 segments form the inner pore of the voltage-gated ion channels came from locating a pore-blocker receptor site—the phenylalkylamine receptor site of L-type Ca^{2+} channels (Striessnig et al 1990). Photoaffinity labelling with high-affinity phenylalkylamine pore blockers showed that only the IVS6 segment of the Ca^{2+} channel $\beta 1$ subunit was labelled. Subsequently, mutagenesis studies of Na^+ channels revealed the local anaesthetic receptor site in an analogous position in the Na^+ channel (Ragsdale et al 1994). High affinity binding of local anaesthetics to the inactivated state of Na^+ channels requires two critical amino acid residues, phe1764 and tyr1771 in brain type IIA channels, which are located on the same side of the IVS6 transmembrane segment two α-helical turns apart (Fig. 5). It is likely that the tertiary amino group of local anaesthetics interacts with phe1764, which is located more deeply in the pore, and that the aromatic moiety of the local anaesthetics interacts with tyr1771, which is

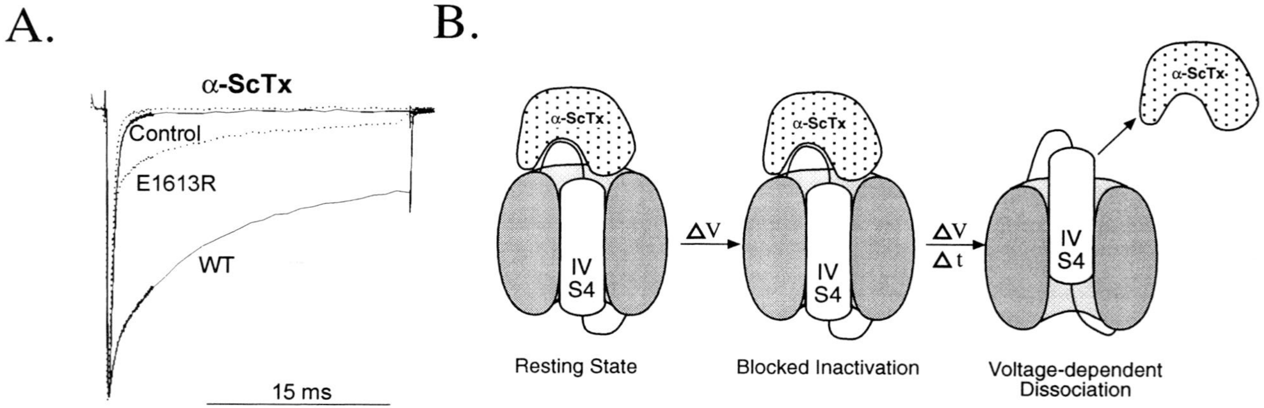

FIG. 4. Voltage sensor trapping by α scorpion toxin. (A) Effect of α scorpion toxin on inactivation of wild-type and mutant Na^+ channels. Na^+ currents were measured at −10 mV by whole-cell voltage clamp in tsA-201 cells transfected with wild-type $Na_v1.2$ channels or mutant E1613R (Rogers et al 1996). Traces with slowed inactivation labelled WT and E1613R were recorded in the presence of 20 nM LqTx, the principal α scorpion toxin of *Leiurus quinquestriatus* (Rogers et al 1996). (B) Voltage sensor trapping model for α scorpion toxin action. Domain IV and its S4 segment are illustrated at resting membrane potential (e.g. −100 mV) and at a depolarized membrane potential (e.g. +40 mV) a short time (e.g. 5 ms) and a long time (e.g. 2 s) after depolarization.

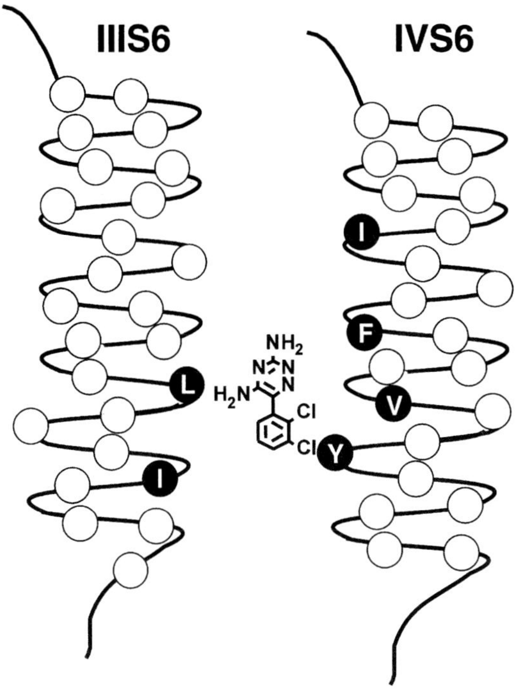

FIG. 5. Lamotrigine binding to the local anaesthetic receptor site in transmembrane segments IIIS6 and IVS6 of the rat brain type IIA Na^+ channel. Side view of the proposed location of the lamotrigine binding site within the pore.

located nearer to the intracellular end of the pore. Recent work has identified two amino acid residues in transmembrane segment IIIS6 that also contribute to the local anaesthetic receptor site (Fig. 5).

Like local anaesthetics, certain anticonvulsant and antiarrhythmic drugs are intracellular pore blockers of Na^+ channels. We have surveyed a selection of these agents to determine whether they also bind at the local anaesthetic receptor site by testing the effect of the mutations F1764A and Y1771A on the affinity for drug binding to the inactivated state. We have found that a wide range of local anaesthetic, anticonvulsant and antiarrhythmic drugs of different structure all interact with these two amino acid residues in the local anaesthetic receptor site. The ratio of effects of these two mutations differs for drugs of different structure, suggesting a specificity of interaction based on drug structure. These results lead to a model in which F1764 and Y1771 are common elements of a complex receptor site for Na^+ channel blocking drugs, which make additional drug-specific contacts that increase their affinity and alter their pharmacological properties.

The common features of the local anaesthetic receptor site suggest that novel Na^+ channel blocking drugs could be developed to fit this site that might have higher affinity and specificity than presently available compounds. Consistent with this idea, the novel compound BIII 890 CL, a complex ether of benzoyl and modified benzomorphan moieties, is a potent Na^+ channel blocker binding at the local anaesthetic receptor site with a K_d of 50 nM for the inactivated state (Carter et al 2000). It is aimed for neuroprotective therapy in the treatment of stroke and possibly neurodegenerative diseases.

Conclusions

Analysis of the structure and function of Na^+ channels has led to many important insights into the molecular basis for channel gating, modification of gating by polypeptide neurotoxins, and block of Na^+ channels by local anaesthetics and related drugs. These studies point to the S4 segments and the S3–S4 loops on their extracellular side as the locus for voltage sensing and for the actions of polypeptide toxins on the gating process. They also point to the S6 segments in domains III and IV as molecular targets for a broad range of Na^+ channel blocking drugs, including local anaesthetics, anticonvulsant drugs, and antiarrhythmic drugs. Targeting these regions for future drug design may yield novel agents that can interrupt hyperexcitability in epilepsy, cardiac arrhythmias and persistent pain syndromes.

Acknowledgements

The research from the author's laboratory was supported by NIH Research Grants NS15751 and NS25704.

References

Armstrong CM 1981 Sodium channels and gating currents. Physiol Rev 61:644–683
Carter AJ, Grauert M, Pschorn U et al 2000 Potent blockade of sodium channels and protection of brain tissue from ischemia by BIII 890 CL. Proc Natl Acad Sci USA 97:4944–4949
Catterall WA 1986 Voltage-dependent gating of sodium channels: correlating structure and function. Trends Neurosci 9:7–10
Cestèle S, Qu Y, Rogers JC, Rochat H, Scheuer T, Catterall WA 1998 Voltage sensor-trapping: enhanced activation of sodium channels by β-scorpion toxin bound to the S3–S4 loop in domain II. Neuron 21:919–931
Cha A, Ruben PC, George AL Jr, Fujimoto E, Bezanilla F 1999 Voltage sensors in domains III and IV, but not I and II, are immobilized by Na^+ channel fast inactivation. Neuron 22:73–87
Chen LQ, Santarelli V, Horn R, Kallen RG 1996 A unique role for the S4 segment of domain 4 in the inactivation of sodium channels. J Gen Physiol 108:549–556
Eaholtz G, Scheuer T, Catterall WA 1994 Restoration of inactivation and block of open sodium channels by an inactivation gate peptide. Neuron 12:1041–1048

Goldin AL, Snutch T, Lubbert H et al 1986 Messenger RNA coding for only the α subunit of the rat brain Na channel is sufficient for expression of functional channels in *Xenopus* oocytes. Proc Natl Acad Sci USA 83:7503–7507

Guy HR, Seetharamulu P 1986 Molecular model of the action potential sodium channel. Proc Natl Acad Sci USA 83:508–512

Hartshorne RP, Catterall WA 1981 Purification of the saxitoxin receptor of the sodium channel from rat brain. Proc Natl Acad Sci USA 78:4620–4624

Hartshorne RP, Messner DJ, Coppersmith JC, Catterall WA 1982 The saxitoxin receptor of the sodium channel from rat brain. Evidence for two nonidentical beta subunits. J Biol Chem 257:13888–13891

Hirschberg B, Rovner A, Lieberman M, Patlak J 1995 Transfer of twelve charges is needed to open skeletal muscle Na^+ channels. J Gen Physiol 106:1053–1068

Horn R 2002 Molecular basis for function in sodium channels. In: Sodium channels and neuronal hyperexcitability. Wiley, Chichester (Novartis Found Symp 241) p 21–33

Isom LL 2002 β subunits: players in neuronal hyperexcitability? In: Sodium channels and neuronal hyperexcitability. Wiley, Chichester (Novartis Found Symp 241) p 124–143

Isom LL, De Jongh KS, Patton DE et al 1992 Primary structure and functional expression of the β1 subunit of the rat brain sodium channel. Science 256:839–842

Isom LL, Ragsdale DS, De Jongh KS et al 1995 Structure and function of the β2 subunit of brain sodium channels, a transmembrane glycoprotein with a CAM motif. Cell 83:433–442

Kellenberger S, Scheuer T, Catterall WA 1996 Movement of the Na^+ channel inactivation gate during inactivation. J Biol Chem 271:30971–30979

Kellenberger S, West JW, Scheuer T, Catterall WA 1997 Molecular analysis of the putative inactivation particle in the inactivation gate of brain type IIA Na^+ channels. J Gen Physiol 109:589–605

Makita N, Bennett PB, George AL Jr 1996 Molecular determinants of β1 subunit-induced gating modulation in voltage-dependent Na^+ channels. J Neurosci 16:7117–7127

McCormick KA, Isom LL, Ragsdale D, Smith D, Scheuer T, Catterall WA 1998 Molecular determinants of Na^+ channel function in the extracellular domain of the β1 subunit. J Biol Chem 273:3954–3962

McCormick KA, Srinivasan J, White K, Scheuer T, Catterall WA 1999 The extracellular domain of the beta1 subunit is both necessary and sufficient for beta1-like modulation of sodium channel gating. J Biol Chem 274:32638–32646

Noda M, Shimizu S, Tanabe T et al 1984 Primary structure of Electrophorus electricus sodium channel deduced from cDNA sequence. Nature 312:121–127

Noda M, Ikeda T, Kayano T et al 1986a Existence of distinct sodium channel messenger RNAs in rat brain. Nature 320:188–192

Noda M, Ikeda T, Suzuki H et al 1986b Expression of functional sodium channels from cloned cDNA. Nature 322:826–828

Qu Y, Rogers JC, Chen SF, McCormick KA, Scheuer T, Catterall WA 1999 Functional roles of the extracellular segments of the sodium channel α subunit in voltage-dependent gating and modulation by β1 subunits. J Biol Chem 274:32647–32654

Ragsdale DS, McPhee JC, Scheuer T, Catterall WA 1994 Molecular determinants of state-dependent block of Na^+ channels by local anesthetics. Science 265:1724–1728

Rogers JC, Qu Y, Tanada TN, Scheuer T, Catterall WA 1996 Molecular determinants of high affinity binding of α-scorpion toxin and sea anemone toxin in the S3–S4 extracellular loop in domain IV of the Na^+ channel α subunit. J Biol Chem 271:15950–15962

Rohl CA, Boeckman FA, Baker C, Scheuer T, Catterall WA, Klevit RE 1999 Solution structure of the sodium channel inactivation gate. Biochemistry 38:855–861

Shapiro L, Doyle JP, Hensley P, Colman DR, Hendrickson WA 1996 Crystal structure of the extracellular domain from P_0, the major structural protein of peripheral nerve myelin. Neuron 17:435–449

Sheets MF, Kyle JW, Kallen RG, Hanck DA 1999 The Na channel voltage sensor associated with inactivation is localized to the external charged residues of domain IV, S4. Biophys J 77:747–757

Srinivasan J, Schachner M, Catterall WA 1998 Interaction of voltage-gated sodium channels with the extracellular matrix molecules tenascin-C and tenascin-R. Proc Natl Acad Sci USA 95:15753–15757

Strichartz GR, Zhou Z, Sinnott C, Khodorova A 2002 Therapeutic concentrations of local anaesthetics unveil the potential role of Na^+ channels in neuropathic pain. In: Sodium channels and neuronal hyperexcitability. Wiley, Chichester (Novartis Fund Symp 241) p 189–205

Striessnig J, Glossmann H, Catterall WA 1990 Identification of a phenylalkylamine binding region within the alpha 1 subunit of skeletal muscle Ca^{2+} channels. Proc Natl Acad Sci USA 87:9108–9112

Stuhmer W, Conti F, Suzuki H et al 1989 Structural parts involved in activation and inactivation of the sodium channel. Nature 339:597–603

Vassilev PM, Scheuer T, Catterall WA 1988 Identification of an intracellular peptide segment involved in sodium channel inactivation. Science 241:1658–1661

Vassilev P, Scheuer T, Catterall WA 1989 Inhibition of inactivation of single sodium channels by a site-directed antibody. Proc Natl Acad Sci USA 86:8147–8151

West JW, Patton DE, Scheuer T, Wang Y, Goldin AL, Catterall WA 1992 A cluster of hydrophobic amino acid residues required for fast Na^+ channel inactivation. Proc Natl Acad Sci USA 89:10910–10914

Xiao ZC, Ragsdale DS, Malhotra JD et al 1999 Tenascin-R is a functional modulator of sodium channel β subunits. J Biol Chem 274:26511–26517

Yang NB, Horn R 1995 Evidence for voltage-dependent S4 movement in sodium channel. Neuron 15:213–218

Yang NB, George AL Jr, Horn R 1996 Molecular basis of charge movement in voltage-gated sodium channels. Neuron 16:113–122

DISCUSSION

Strichartz: One of the interesting things about the high-affinity inactivated state binding is that if we posit that contributions are made from all four S6 helices, then it is odd that when we mutate one amino acid in one of the helices there are huge effects on affinity. You might think that you'd lose a third or a quarter of the binding energy, but in fact it looks like the affinity is near zero.

Catterall: The effects aren't that large. Binding energy is a logarithmic function, so a 10-fold change in binding constant results from a much smaller change in binding energy. I think it is consistent with the idea that there are multiple sites of attachment. Another correlation which is not strictly quantitative is that the larger the drug, the less impact mutating the two residues in the IV S6 segment has on its binding. Both those mutations have a 300-fold effect on lidocaine, but if you do the same experiment on quinidine or flecainide, which are much bulkier drugs, they have more like three- or fivefold effects. One explanation for this is that the large drugs touch more places, so when you mutate two residues it has less impact. Our thinking is that the mutations we make are only reducing a portion of the binding energy, and this portion is greatest for the drugs of simplest structure, because they make the fewest contacts with the pore.

Strichartz: An alternative explanation is that these mutations affect the conformation of the channel and indirectly affect the binding site. So when we change them, the channel will still inactivate, but it folds up into a tighter or a looser situation, so this binding site, which has a weak stereoselectivity to begin with, is altered.

Catterall: That is certainly possible. It is almost impossible in a mutagenesis experiment to rule out conformational effects, but it is important to say that in these experiments there is almost no correlation between the effects of mutations on inactivation and their effects on drug block. You can have large effects on inactivation and no effect on drug block, and vice versa. It is not likely that all of the effect here is conformational, simply preventing the inactivated state from forming.

Strichartz: It is not a question of preventing it from forming, but rather of the actual form of the inactivated state.

Catterall: There is no way that we can test that without a crystal structure. For Na^+ channels this is still a few years away.

Cummins: I have been told that one of the problems for using these drugs for neuropathic pain is that they are just not specific to peripheral ion channels. There is evidence in the literature, however, that the SNS-type current does have lower affinity to some of these agents, although not all. Does this indicate that we will be able to target this area with subtype-specific drugs?

Catterall: It is a possibility, but the specific residues in the IV S6 segment we have identified are present in SNS, so it isn't those residues that are causing the difference. It might be residues in other S6 segments. I haven't gone back to look at the other residues we have identified in domain III and domain I to see whether they are present in SNS. There is clearly an affinity difference. It might be that this is due to the slower and weaker inactivation of those channels, and it is not a receptor site phenomenon but instead a conformational change/allosteric-type phenomenon; that is, that the inactivated state is different for those channels and does not provide such an appropriate binding site for these drugs. This again might argue that you could design a drug that would be specific for these channels and would target this site. I am sure that many drug companies are trying to do this; I think it is a reasonable hypothesis. I think it is encouraging that Boehringer–Ingleheim was able to make a Na^+ channel-blocking drug with a K_d of 50 nM. We don't have any other Na^+ channel-blocking drugs that act at 50 nM concentrations. Their therapeutic target for this is neuroprotection, and they have this drug in stroke trials at present.

Horn: You talked about a couple of different classes of compounds that block the pore. One of them was a local anaesthetic and the other was a peptide. Have you ever looked at whether the affinity of this peptide is changed by your S6 mutations?

Catterall: It is not affected by the F1764A or Y1771A mutations, so we think that the binding site for the inactivation gate receptor is different. I suppose the crystal structure of the K_{csa} channel I showed can illustrate the point. We think the drug binds in what I called the cavity of the structure, but the inactivation gate can't get there. It has to bind more superficially to that helix cross near the intracellular end of the S6 segments, where the amino acid residues that we and Al Goldin have found to be essential for inactivation may be located in the Na^+ channel structure.

Horn: When Galen Eaholtz was studying that, she looked for a voltage dependence to the block. As I remember, she found some voltage dependence for the peptide reaching its binding site, but it was very superficial.

Catterall: Remember, the charges in the KIFMK peptide are not in the natural sequence, and therefore they are probably not binding to the receptor site. We think that the charged peptide is being concentrated in the diffuse double layer by the mechanism that Gary Strichartz suggested, and that you see voltage dependence and a requirement for charge in order to get enough peptide near the mouth of the pore to give block during the millisecond time frame that the Na^+ channel is open.

Horn: I don't see how that is going to give a voltage dependent block. I can understand that the fact that you have lysines on either end of it might concentrate the peptide there, but in order to sense the electric field it has to get into the electric field. If it is right on the very mouth of the channel it is not going to feel the electric field gap.

Catterall: There are two components. You get voltage-dependent block because the channel needs to form the inactivation gate receptor, and then there is voltage dependence to do with the concentration of the drug. In addition, Eaholtz and I found a second form of block by IFM-containing peptides. This second form of block was strongly voltage dependent and slowly reversible, as if the peptide entered more deeply into the pore.

Horn: I had the impression that they found a voltage dependence that looked like open channel block voltage dependence. They had a channel that didn't inactivate and they could see it as a charged open channel blocker entering the electric field. It seems to me that there is some inconsistency here: I don't see how it is entering the electric field if it is at the very mouth of the channel.

Strichartz: If the field potential changes when the blocking particle is there, because now no ions can pass through the pore, a voltage dependence will result from the blockade per se. It will show up in the off-rate constant.

Horn: You mean that it is there already.

Strichartz: The electric field across the membrane has to drop over the same potential, from bulk outside to bulk inside. If, when a blocking particle is in place, the profile of ion mobility is altered, then the profile of the field across the channel is also different, which can give a voltage dependence that would show up in the off-rate constant, but not in the on-rate constant.

Horn: So the field is actually dropping across the blocker.

Catterall: And the field drops across that region differently when the channel is open and closed. Again, if the blocker is in that region it is going to feel that change in electrical potential.

Waxman: Earlier Bill Catterall mentioned neuroprotection. In that context, a few years ago Peter Stys, Bruce Ransom and I looked at the issue of Ca^{2+}-mediated injury in white matter, where there are no synapses (Stys et al 1992a). Not surprisingly, glutamate doesn't trigger Ca^{2+}-mediated injury in white matter. Peter did identify a pathway involving Na^+ influx via a persistent Na^+ conductance which depolarizes axons and also reverses the transmembrane Na^+ gradient and drives that Na^+/Ca^{2+} exchanger into a reverse node (Stys et al 1992a). This is the death knell of axons within anoxic white matter. The interesting thing was that one could protect white matter axons from this cascade with both tertiary anaesthetics (procaine and lidocaine), and quaternary anaesthetics (QX314 and QX222). This protection occurred at concentrations that did not impair impulse conduction (Stys et al 1992b). One final point was that it may have been fortuitous that some of these compounds weren't clean: they depressed P_K as well as P_{Na}, and this may have helped to preserve conduction.

Catterall: I suppose they are acting to preferentially block the long opening Na^+ channels and do not have much effect on those open at the peak of the spike.

Bean: What is your view of the origin of the change in affinity between the resting and inactivated states? Do you view that as being a change in the binding site? It couldn't just be a change in accessibility of the binding site when the channel opens.

Catterall: I imagine it as a conformational change in the binding site. We have one indirect bit of evidence for that. We found that in the III S6 segment, the periodic effect on activation when we make those alanine substitution mutations has the exact periodicity of an α helix. It can be a pretty strong effect (20 mV shifts in activation curve). It is as if one face of the helix makes interactions in the closed state that are different from those it makes in the open state. We think this helix must be turning during the activation process and exposing a new face to either interactions with solvent or interactions with other transmembrane segments, giving us this periodic pattern of effects on activation. We don't see the same periodic pattern in the IV S6 segment, but it may be that all of them rotate a little bit. Our model is that these segments rotate and open the pore, sort of like opening a camera lens, and in doing so expose phenylalanine 1764 and tyrosine 1761 to the lumen of the pore, where these drugs can then bind to them. This is why those two residues are common determinants of inactivated state binding for all the drugs, because they become accessible to the drug binding site during the channel opening process and inactivation.

Bean: Is it an open possibility that the high affinity binding is a consequence of the activation of the channel — the movements of the S4 regions — and doesn't necessarily result from the inactivated state? It could be that the movement of the S4s produces a high affinity site for the local anaesthetics, and also promotes formation of the inactivated state.

Catterall: I think that is right, but there is another way to think about this. For most of these drugs, the affinity for the inactivated state is higher. Activation causes rotational movements of helices that reveal the binding site, even in the open state of the channel. The drugs enter and bind. But then when the channel inactivates, the four S6 segments move a little closer, so the drug can bind to all four domains at the same time, or to three of them, rather than just one. So the structure collapses a little bit when you reach the inactivated state. This would be a model that would allow us to understand higher affinity binding to the open state and even higher affinity binding to the inactivated state.

Bean: Can you see a difference in binding if you use the inactivation gate mutants to prevent inactivation?

Catterall: There are some different results in the literature, but we haven't looked at this systematically. If we make just the phenylalanine 1489 mutation, which is sufficient to substantially block inactivation we still observe pretty good use-dependent block by local anaesthetics. Other groups have mutated all three residues of the IFM motif and saw a much bigger loss of local anaesthetic binding, so they concluded that the gate was critical to the receptor site. We don't find that when we just use a single mutation. Perhaps the triple mutation has conformational effects beyond just preventing the closure of the inactivation gate.

Strichartz: Moving to peptide toxins for a moment, years ago we did experiments looking at the effect of combining α toxins and β toxins on nodal Na^+ currents, and saw that there was a marked synergy, particularly in terms of enhancing and prolonging the slowly declining current induced by β toxins (Wang & Strichartz 1982). Looking at your model, this would suggest that there is some coupling between S4s in domain II and domain IV. Either that, or the two toxins at the surface somehow interact.

Catterall: Aren't the scorpions clever! They have one peptide toxin pulling out one S4 segment, and the other pushing the other S4 segment in and keeping the channel in an open state.

Strichartz: And then they add K^+ channel blockers to make it even worse!

Isom: A while ago, we had the idea that the binding site for α scorpion toxin was the site of interaction between $\beta 1$ and α. Do you still think that is true?

Catterall: No. In photoaffinity experiments we observed labelling of both α and $\beta1$ subunits. We can understand this now, because it is probably a bystander effect for the $\beta1$ subunit. The scorpion toxin receptor site is certainly not on the $\beta1$ subunit, and Na^+ channels that have only an α subunit are fully sensitive to scorpion toxin. $\beta1$ must be near enough to the α scorpion toxin binding site to be photolabelled as a bystander. One point of interaction of $\beta1$ is the last extracellular loop, just outside the IV S6 segment. This is the extracellular loop exactly adjacent to the one that α-scorpion toxin binds to, which is the S3/S4 loop in domain IV. We think the toxin sits there, the $\beta1$ subunit is nearby and sometimes the photoprobe reaches over and covalently labels it.

Segal: You have made a compelling case for the local anaesthetics and anticonvulsants working at the same site. Do you have any insights as to what would account for differences in types of actions between those groups?

Catterall: I think the correlation is that anticonvulsant drugs have voltage-dependent binding and they are able to bind to the inactivated state, even if the channel is not opened by strong depolarization. They have a strong component of voltage-dependent block and a little bit of frequency-dependent block superimposed on that. This is true of phenytoin and lamotrigine, for sure. In contrast, the anti-arrhythmic drugs tend to have very strong frequency dependence, as lidocaine does, along with some voltage-dependent block. I think it is a ratio of how well they bind to the open state, compared with the inactivated state. The drugs that are good anticonvulsants bind preferentially to the inactivated state because they are probably very effective at blocking the sustained Na^+ currents.

Wood: Do you have any insight into the structural basis of μ conotoxin binding to Na^+ channels?

Catterall: We haven't worked on this, but others have. Harry Fozzard's group at the University of Chicago have produced a detailed model for how both tetrodoxin and the μ conotoxins might bind to the extracellular mouth of the pore of Na^+ channels. I can't cite the residues, but their papers lay out a detailed model supported by mutagenesis experiments.

Cummins: One of the problems with that model is that most of the residues they have identified are conserved in neuronal and skeletal muscle channels. They don't really get to the idea of why the conotoxins mainly hit the skeletal muscle channels. We recently followed up on an observation that Dick Horn made a couple of years ago about the fact that the human skeletal muscle channel is much less sensitive than the rat skeletal muscle channel. It appears that residues further in the extrapore region, not in the pore but out on the linker, are much more important in subtype specificity.

Goldin: The model for the α scorpion toxins is that they prevent the domain IV S4 helix from moving out a little bit. If this is true, either you should see a slowing of activation, or domain IV S4 doesn't have to move to open the channel.

Catterall: I like the model in which IV S4 has to move in steps. The first steps are important for activation, and they aren't much affected by scorpion toxin binding. Then there are second steps which are slower and important for inactivation. These steps are interfered with by scorpion toxin binding. One of the predictions of this model is that the gating current should be affected by these toxins. There are data to show this, some of them quite old. In the 1980s, Hans Meves showed that β scorpion toxins block gating currents (Meves et al 1987) and Wolfgang Nonner showed that α scorpion toxins block a slow component of gating current (Nonner 1979). More recently, Dorothy Hanck followed this up in a series of experiments that showed that α scorpion toxins prevent gating current due to the outer-most charge (Arg residue) in the IV S4 segment (Sheets et al 1999). If this is neutralized, then the component of gating current that is blocked by scorpion toxins is largely lost. All of this agrees quite well with the idea that IV S4 voltage sensor is rather specific for inactivation (e.g. this is how scorpion toxin affects inactivation), and that this S4 segment really has two jobs: a secondary role in activation and a primary role in inactivation.

Keynes: In the model that Elinder and I produced a few years ago, based on limited evidence, the difference between the inactivated states and the non-inactivated state was simply a hydration step. Does what you are saying fit with this?

Catterall: I don't think it argues against it, but the evidence that the IV S6 segment needs to move in order for inactivation to occur is fairly strong. This movement may alter hydration of those gating charges. The two may be synonymous. If you move a gating charge into a hydrophilic environment, then hydration will change. I wouldn't be surprised if both things occur.

Horn: The inactivation gate itself is going from a hydrophilic environment to a hydrophobic one. Your data showed that the cysteine that is substituted for phenylalanine becomes inaccessible. Presumably it is up against some protein surface excluding water, so there has to be a change of hydration.

References

Meves H, Rubly N, Watt DD 1987 Gating current experiments on frog nodes of Ranvier treated with *Centruroides sculpturatus* toxins or aconitine. Pflügers Arch 409:381–393

Nonner W 1979 Effects of *Leiurus* scorpion venom on the 'gating' current in myelinated nerve. Adv Cytopharmacol 3:345–352

Sheets MF, Kyle JW, Kallen RG, Hanck DA 1999 The Na channel voltage sensor associated with inactivation is localized to the external charged residues of domain IV, S4. Biophys J 77:747–757

Stys PK, Waxman SG, Ransom BR 1992a Ionic mechanisms of anoxic injury in mammalian CNS white matter: role of Na^+ channels and Na^+–Ca^{2+} exchanger. J Neurosci 12:430–439

Stys PK, Ransom BR, Waxman SG 1992b Tertiary and quaternary local anesthetics protect CNS white matter from anoxic injury at concentrations that do not block excitability. J Neurophysiol 67:236–240

Wang GK, Strichartz GR 1982 Simultaneous modifications of sodium channel gating by two scorpion toxins. Biophys J 40:175–179

Final general discussion

Cummins: I want to make a general comment, which follows up on an issue raised by Indira Raman earlier in the meeting. Persistent currents have been implicated in a lot of the pathophysiologies. For example, slowly inactivating currents are important in paramyatonia congenita, window currents are important in hyperkalaemic periodic paralysis, and non-inactivated currents are probably important in epilepsy. One of the problems with this is that not all persistent currents are created equally, and they are often not all defined equally. What one person shows as non-inactivated on a 20 ms time-scale may be clearly slowly inactivating on a 500 ms time-scale. I think we need to be more rigorous in how we describe the persistent currents. This is probably going to be harder than getting the molecular biologists to agree on terminology for the different channel isoforms, but at least when we describe a persistent current, we should be very careful to say what we mean by that. The therapeutic strategies to tackle different persistent currents may be very distinct, depending on whether it is created by modulation or mutation or some other mechanism.

Raman: Another relevant issue is whether the window currents, or currents measured at sub threshold potentials, come from channels that will inactivate at more positive potentials. This issue becomes relevant when we start talking about currents that participate in rapid firing. For instance, if there is a persistent current that is present at −60 mV, but the channel producing this current will inactivate at more positive potentials, it is harder to implicate it in very rapid firing, but such a current may be very important in slow depolarizing potentials that lead to pacemaking. I think this is another component that is important in addition to how long after a step depolarization we measure the current.

Horn: Another issue that has hardly been raised in this meeting is that some of the ion channelopathies are due to changes in slow inactivation, which is a process quite different from what we have been talking about. The origin of a persistent slow current could be a different isoform of the channel, or it could come from different processes within the same channel.

Cummins: There again, we run into the problem of what is 'slow inactivation'? The SNS current is described as a 'slowly inactivating' current. It has slow inactivation, but it is not the same as slow inactivation that is altered in the periodic paralysis mutations.

Bean: I think part of the issue has been that the term 'persistent' has evolved a bit. When Llinás introduced the term in his studies looking at action potentials in Purkinje neurons, he was thinking of a completely separate class of Na^+ channels which would have no inactivation at all: they would just activate. I think it's fair to say that since then, as Wayne Crill and others have looked at the channels under voltage clamp, it is uncommon to see this behaviour. More likely, the not-inactivating component is coming from channels which can or do inactivate by 80–95%. I don't know what this means in regard to how we should use the term. I don't think many of us think of a separate class of channels which don't inactivate at all.

Crill: Several years ago there was a single abstract reporting a separate channel without inactivation properties in cerebellar Purkinje cells. No full report has been published. The term 'persistent' was used to describe the channel that stayed open, and at least at that time no one had looked at inactivation properties. I think they probably do inactivate slowly, and as far as I am concerned no one has yet showed a separate channel that doesn't inactivate, and almost everyone thinks it is the behaviour of regular Na^+ channels.

Raman: Yet in modelling studies we continue to see a separate I_{NaP} that is defined as a kinetically distinct entity, and doesn't seem to follow from a steady state component of a channel which has certain rate constants between closed, open and inactivated states. After this discussion, it seems that most of us would assert that the majority of the long-lasting current that we measure doesn't come from a separate channel, yet it continues to show up in textbooks, conversation and modelling.

Strichartz: Speaking of modelling, when we want to address the contribution of altered channel gating to observed pathologies, when we take biophysical data which we must generate under less than physiological conditions, we have to be very cautious in extending this to understand sustained or repetitive firing. As Dick Horn mentioned, with cell-attached patches, this disturbs the membrane by sucking it up into the pipette. When you use good buffers such as HEPES instead of bicarbonate, you effect a shift of some 10 mV in inactivation behaviour. None of us is going to work at 37 °C: it would be all over before the capacitative surge had settled. We have to be cautious in extending our biophysical data into a domain to explain pathophysiology.

Spruston: I think it is possible to work at 37 °C: we now do all our work at this temperature. Returning to slow inactivation, another important point is to distinguish between the rate of entry into the inactivated state and the rate of recovery from it. We find that in hippocampal neurons, slow inactivation is actually slow recovery from inactivation as opposed to slow entry. This is particularly important, because the most efficient way to drive channels into that state is with repetitive brief depolarizations in voltage clamp experiments. In the physiological context this is very relevant, because during rapid, sustained, action potential firing you are going to get much more of this so-called slow inactivation than you get with long depolarizing pre-pulses in voltage clamp experiments. But we think the same state is entered with either type of depolarization. You can enter this state either directly from the open state, or through the fast inactivated state. Also, the rate of entry into the slow inactivated state is much more rapid from the open state. With long pre-pulses you drive most of the channels rapidly to the fast inactivated state, and then you can slowly trickle into the slow inactivated state. With repetitive depolarizations you can keep going, looping through the open state, and each time a small fraction rapidly enters the slow inactivated state, which then takes in the order of one second at the resting potential to recover. It is important to think carefully about the rate of entry and rate of recovery from slow inactivation, and how we define this slow inactivation (see the state diagram in Mickus et al 1999).

Waxman: Ted Cummins, you have thought a lot about the recovery from inactivation. Do you have anything to add?

Cummins: To follow up on this, it is certainly not the same for all the isoforms. With the skeletal muscle channel you can knock out much of the fast inactivation with the F1304Q mutation and it doesn't affect the biophysically distinct slow inactivation, either the entry or the exit. This doesn't mean that this isn't the case for some of the neuronal channels.

Spruston: I didn't mean to say that slow and fast inactivation are the same.

Cummins: But your data may imply that they are coupled. The coupling is not that great for the skeletal muscle channel, although Featherstone et al (1996) reported that there is some. There are also differences in repriming and entry into the fast inactivated state between many of the different isoforms. For the PN1 channel those kinetics are very slow, and for the skeletal muscle channel they are fast. Now we are seeing that these types of transitions are potentially important for epilepsy. We tend to generalize over Na^+ channels, and the modellers have done this to the extreme, where there is one fast Na^+ current that is the same in all the neurons. It is probably not. The Na6 isoform clearly has different kinetics than the PN1, and you can't model a neuron that has both isoforms and get the real behaviour using just one type of channel.

Spruston: It's clear that different isoforms have different properties. This is an important area for future investigations. Those of us who are interested in the CNS and specialization of firing properties would like to know more about how different isoforms have an effect. Another really important factor may be the interaction of the Na^+ channels with the cytoskeleton itself, through things like AKAPs (A-kinase anchoring proteins) and with extracellular elements through the β subunits, and how this affects Na^+ channel properties. Clearly, every cell in the CNS is different, and the Na^+ channels are rarely alike in any two types of neurons.

Meisler: On the subject of isoform differences and potential therapeutic targets, we haven't heard anything about the cytoplasmic loop domains. We usually emphasize the divergence of these domains in different Na^+ channels, e.g. 1.2 versus 1.8. However, the cytoplasmic loops of each type of channel are very highly conserved between different species. For example, the loop domains of SCN1A are more than 98% conserved between human and rodent, suggesting that they have a very important function, or else they would be diverging must faster. The same is true for $Na_v1.6$ (SCN8A). This seems to offer a potential target for isoform-specific interventions. Are there any good functional data for the loop domains, besides their phosphorylation?

Catterall: One of the things that clearly occur in the large intracellular domains is modulation. There are phosphorylation sites. These are isoform specific. Skeletal muscle channels are not modulated by PKA, for example, and Al Goldin mentioned earlier that perhaps $Na_v1.6$ is not strongly modulated by PKA either. Somewhere there is a G protein $\beta\gamma$ binding site, presumably on one of the intracellular loops. I think is an important and underappreciated aspect of Na^+ channel biology; that is, they are modulated in just the way that we know that Ca^{2+} and K^+ channels are.

Keynes: I have a very general question about the evolution of Na^+ channels. How much information is there? Hille has said that he thinks that Ca^{2+} channels developed from K^+ channels, and then Na^+ channels came after them. I am surprised by how many of the different subspecies of Na^+ channels occur in *Drosophila*.

Catterall: There is some information from the genome projects. In *Drosophila* there are just two Na^+ channels in the entire genome. In *Caenorhabditis elegans* there is no sequence that looks exactly like a Na^+ channel, although there is one that might be. *C. elegans* has three types of Ca^{2+} channel, which are similar in structure to the three families that are present in vertebrates. In vertebrates there is the Ca_v1 family (which makes L-type Ca^{2+} currents), the Ca_v2 family (which makes N, P, Q and R-type Ca^{2+} currents) and the Ca_v3 family (which makes T-type Ca^{2+} currents): there is one of each of these in *C. elegans*.

Noebels: It is said there are more than 50 types of K^+ channels in *C. elegans*.

Bean: Are there Na^+ channel β subunits in *Drosophila*?

Isom: No. There is TipE, which is an auxiliary subunit, but it doesn't look like the β subunits in terms of amino acid sequence or protein structure. The other point about Na^+ channels in *Drosophila* is that although there are only two genes, there are many splice variants.

Waxman: Miriam Meisler, you have thought about evolution and reduplication. Do you have any comments?

Meisler: We can only recognize orthologues within the mammals, because we know the chromosomal relationships. In the fish there have been some extra chromosome duplications resulting in extra copies of channel genes that do not correspond to any specific mammalian gene.

Noebels: Plants are interesting, too. They have ion channels.

Strichartz: They have proton channels and chloride channels, and all organisms have K^+ channels. Although excitability has been reported in *Nitella* and there are action potentials in roots, they are not Na^+ action potentials. You can get a regenerative phenomenon any time you get a potential dependent current that leads to a change in the direction that continues to increase the permeability.

Segal: It sounds like we need Na^+ channels in order to be fast, and some of the diseases we are discussing may be consequences of life in the fast lane.

Horn: There is certainly strong evolutionary pressure for being able to respond quickly to things.

Bostock: I have a question about Na^+ channels and persistent currents. I have been surprised that I have found so little in the literature about the effects of pH on these. Mark Baker and I found that persistent Na^+ currents in large dorsal root ganglion (DRG) neurons are particularly pH sensitive (Baker & Bostock 1999). They seem to be about half blocked at normal pH. We related this to the induction of paraesthesiae by hyperventilation or metabolic alkalosis (Mogyoros et al 1997, 2000). The paraesthesiae are due to spontaneous discharges, which can be induced rapidly in cutaneous afferents with small pH changes in the absence of changes in serum ionized Ca^{2+}. Michael Segal's persistent currents in cortical cells underlying epilepsy look remarkably like some of the persistent currents we recorded in DRG cells. Hyperventilation has been used as a provocative manoeuvre to induce spike and wave EEG activity as a test for epilepsy. My question is, are cortical and other persistent Na^+ currents also very pH sensitive, or is this restricted to particular isoforms? Is this relevant to the skeletal muscle channelopathies, for which we have heard that carbonic anhydrase inhibitors provide a general, non-specific treatment? Could it be relevant to the mechanism of the ketogenic diet, a non-specific treatment for childhood epilepsy, originally introduced to mimic the anticonvulsant effect of fasting (which produces ketosis and acidosis) (Stafstrom & Spencer 2000)?

Segal: Clinically, hyperventilation is used to elicit the absence-type of seizure, which depends on a thalamic circuit that predominantly involves Ca^{2+} currents.

Strichartz: CO_2 has general anaesthetic properties: at high pressures it drives things in the other direction, hypo-excitability.

Segal: I haven't tried hyperventilation on people with classic cortical seizures — I haven't heard of it having a substantial effect. It has an impressive and rapid effect on people with thalamic type of seizures.

Crill: Isn't it thought that in hyperkalaemic periodic paralysis one explanation for the ability to prevent the onset of paralysis is pH changes.

Ptacek: That is thought to be the case. It isn't clear, however, whether it is through action on the Na^+ channel itself. In muscle CLC1 in the Cl^- channel is extremely sensitive to pH. It could be acting on some other channel, but many people think it is pH. Supporting this in the more generalized way is the fact that carbonic anhydrase inhibitors help some patients with migraine. Topiramate has some carbonic anhydrase inhibition properties, although it is not clear that these play any role in its anticonvulsant effects.

Strichartz: As I mentioned before, bicarbonate alone, under controlled pH conditions (of course, you have to let CO_2 change), will affect Na^+ channel inactivation. Increased bicarbonate ions enhance inactivation. It may be that the change in the CO_2/bicarbonate balance and not pH per se (which may be better buffered *in vivo*) is exerting this effect. With respect to carbonic anhydrase inhibitors, Steve Connor's graduate student, Donella Green, did a thesis looking at Diamox (acetazolamide), a classic CA inhibitor, and showed that it had use-dependent blocking properties on muscle Na^+ channels.

Bean: It is unclear how general such direct effects of acetazolamide are. Donella Green came over to our lab and we did some experiments together on Na^+ channels in neurons, and didn't see significant blocking activity. It may be unclear whether in her experiments on skeletal muscle channels it was a direct effect on the channels or something less direct on the cells. It looked unlikely to her and me that there was a direct effect on neuronal Na^+ channels.

Segal: If Na^+ channels are needed for life in the fast lane, it seems ironic that the diseases we are discussing are ones not of the peak Na^+ current, but of the persistent Na^+ current. This raises the question of what is it about the persistent Na^+ current that causes it to be present in so many different cell types; what is the selective advantage of the persistent Na^+ current? Is the persistent current just a kinetic abnormality that is a loose end that has not been eliminated by evolution, or is there an advantage to this persistent Na^+ current? It would seem to be important not only philosophically to gain a better understanding of the advantages of the persistent Na^+ current, but also if we are to design drugs that specifically reduce the persistent Na^+ current, it would be good to know what types of functions that intervention might affect in addition to the disease states we want to treat.

Bean: I can echo Rodolfo Llinás' ideas on this, which I think are insightful. A primary result from persistent Na^+ current is to induce overall oscillatory activity in the CNS. This is crucial for the function of the nervous system in some way we still don't understand. A cost of this is the possibility of going into a less well controlled oscillatory activity. The normal oscillatory activity that you see reflected in the normal EEG may well reflect the activity of the persistent Na^+ currents in some cell types.

Strichartz: If you have a membrane with a low density of channels (and the critical ratio here is the density of channels over the specific membrane capacitance), then you'll have a much greater effect of a channel that stays open for longer. During an action potential in a fast-conducting fibre, about 20% of the channels actually open during that spike. This is why you have such a wonderful margin of safety. The spike rises more slowly under various toxic conditions, which allows more of the channels to open. You have to be really sick to get conduction failure. This behaviour is largely due to the capacitative properties of the membrane. A channel that is only open transiently will have a modest effect on a large area when channels are present at low density, but a channel that can open repeatedly and continue to supply charge, will have a much larger effect in influencing the total membrane potential. One might find various teleological arguments around that electrophysiological factor to rationalize persistent Na^+ currents.

Cummins: We have to be careful not to generalize when talking about persistent currents. Persistent currents could result, for example, from G protein modulation, PKC modulation or from modal gating without any underlying modulation. There are probably multiple persistent currents just as there are multiple fast Na^+ currents. The physiological consequences of different persistent currents are going to depend on the different biophysical properties of these currents.

Waxman: I would not be surprised to see that, if we were to reconvene this meeting five years from now, we will have identified three or four (maybe more) Na^+ channels that can generate persistent currents. And these different persistent currents will have different physiological properties and different functional roles.

I have been told that our time is up. Again, I would like to thank the Novartis Foundation for its hospitality and superb organization. I would like to thank the Yale–London Collaboration for its encouragement in organizing this meeting. And I would like, most of all, to thank the speakers and discussants.

References

Baker MD, Bostock H 1999 The pH dependence of late sodium current in large sensory neurons. Neuroscience 92:1119–1130

Featherstone DE, Richmond JE, Ruben PC 1996 Interaction between fast and slow inactivation in Skm1 sodium channels. Biophys J 71:3098–3109

Mickus T, Jung H, Spruston N 1999 Properties of slow, cumulative sodium channel inactivation in rat hippocampal CA1 pyramidal neurons. Biophys J 76:846–860

Mogyoros I, Kiernan MC, Burke D, Bostock H 1997 Excitability changes in human sensory and motor axons during hyperventilation and ischaemia. Brain 120:317–325

Mogyoros I, Bostock H, Burke D 2000 Mechanisms of paresthesias arising from healthy axons. Muscle Nerve 23:310–320

Stafstrom CE, Spencer S 2000 The ketogenic diet: a therapy in search of an explanation. Neurology 54:282–283

Index of contributors

Non-participating co-authors are indicated by asterisks. Entries in bold type indicate papers; other entries refer to discussion contributions.

Subject index

D

E

F

G

Q

R

S

T